Forgotten ways for
MODERN DAYS

Forgotten ways for
MODERN DAYS

Kitchen cures and household lore
for a natural home and garden

RACHELLE BLONDEL

photography by CATHERINE GRATWICKE

...LE BOOKS

contents

foreword 6

introduction 8

house & HOME 10

CLEANING 12
wood wipes

beeswax furniture polish

spring clean bucket

forest-fresh floor cleaner

all-purpose orange cleaner

cleaning without chemicals

cleaning glass

the wonder of lemon

LAUNDRY 26
household stain powder

stains

top tips for keeping your whites white

wool dryer balls

wire laundry basket

fabric fresh spray

KITCHEN 36
market day tote bag

reusable beeswax food wrap

jar & bottle care

willow wire herb hanger

harvest apron

cotton produce bags

AROUND THE HOUSE 52
homemade paint

bits & bobs eiderdown

candle in a jar

top tips for saving energy

homemade smudge sticks

fabric backed blankets

kindling wraps

the wonder of eggshells

homemade reed diffuser

in the GARDEN 68

herbs
nourish your soil naturally
garden friends
windowsill salad
plant tonic
compost booster
decal plant markers
what's in season
wild food
how to help bees
the wonder of honey
bird-feeder bottle
keeping chickens

natural HEALTH 90

vital vinegar
homemade apple cider vinegar
cures from the kitchen
garlicky goodness
herbal honey
dandelion syrup
ginger compress
rejuvenating borage & honey tea
honey & lemon soother
keep well syrup
caraway tea for upset stomach
lemon balm ointment
fragrant bath teas
buttermilk wrap for sunburn

natural BEAUTY 114

almond & rosewater cleansing cream
lavender
banish acne tonic
honey soap
sunny sunflower seed mask
solid beeswax perfume
homemade infused essential oils
lavender & cider vinegar toner
care when using essential oils
nettle & chamomile hair rinse
herbal hair rinses
egg hair treatment
leather nail buffer
beeswax nail cream

suppliers and interesting information 139
index 140
acknowledgements 143

foreword

DEAR READER,

THE BOOK YOU HOLD IN YOUR MITTS HAS BECOME A 'BIBLE
OF SORTS' FOR ME AND I AM HOPEFUL UPON READING, IT WILL
BECOME THE SAME FOR YOU. I HOPE YOU WILL AGREE THAT
MY CRAFTY BUDDY RACHELLE HAS A RATHER MAGICAL WAY
OF TAKING THE ORDINARY AND MAKING IT EXTRA-ORDINARY.
BEFORE YOU KNOW IT, 'FORGOTTEN WAYS FOR MODERN DAYS'
WILL HAVE YOU BECOMING MINDFUL IN THE EVERYDAY AND TAKING
TIME OUT FROM THE STRESS AND MESS OF THE CONVEYOR BELT
OF LIFE TO BEING PRESENT IN THE MOMENT.

THIS BOOK SHARES MANY WAYS TO MAKE TIME STAND STILL
A LITTLE, TO BREATHE IN AND SEE THE BEAUTY THAT SURROUNDS
US IN NATURE AND HELP US GET BACK IN TUNE WITH OUR MINDS,
OUR BODIES AND OUR LITTLE LIGHT.

WISDOM AND PRACTICES TAKEN FROM THE PAST HAVE THE
ABILITY TO TEACH US HOW TO TAKE CARE OF OURSELVES AND
OUR PLANET IN THE HERE AND NOW, IF WE CHOOSE TO LISTEN.

BY MAKING EVERYDAY RITUALS SUCH AS CARING FOR OUR

HOME, OUR BODIES AND OUR SURROUNDINGS IMPORTANT

AGAIN, WE ARE ABLE TO FIND JOY AND GRATITUDE.

AFTER TRYING OUT A FEW OF THE RECIPES, I HAVE

NO DOUBT ALONGSIDE OF RACHELLE'S KNOWLEDGE AND

KNOW-HOW YOU WILL SOON BE EXPERIMENTING WITH

POTIONS, LOTIONS AND CREATIONS OF YOUR OWN MAKING,

WITH YOUR HANDS AND HEART. TRULY IF YOU ARE ON THE

PATH TO FINDING A SLOWER AND SIMPLER WAY OF LIVING

YOUR LIFE, I BELIEVE YOU CANNOT GO WRONG WITH ADDING

THIS WISE LITTLE BOOK TO YOUR 'BIBLE OF SORTS' BOOK SHELF.

TIF FUSSELL (DOTTIE ANGEL)

introduction

THIS BOOK GIVES A NOD AND A WAVE TO THOSE HOMEKEEPERS, GARDENERS, CRAFTERS AND KITCHEN ALCHEMISTS WHO TROD THIS PATH BEFORE US. Without access to modern-day bells-and-whistles, solutions for keeping homes clean and gardens in order were found from natural products and things that were close to hand. Ingredients from kitchen cupboards, picked from the garden, or foraged from hedgerows were used to clean the house, cure a chesty cough, freshen the skin or whip up a woolly scarf.

Some of the knowledge and wisdom, along with a pinch of love, has been passed on in this book, to bring the crafty ways of those ingenious homemakers into modern times. More often than not, our grandparents were forced, out of necessity, to be creative and use what was to hand, whereas now we are bombarded with 'stuff' that we 'cannot possibly do without'. Many of us just want to get back to basics and live a simpler life. With this book, I aim to pass on information to help you do that.

Bringing simplicity and natural products into your everyday life is much easier than you'd imagine, and the joy of concocting various recipes can make a dull task seem a little cheerier. It's rewarding to use the things you make to ease and enrich your life, and in doing so you produce less waste too. In this book, I show you how to make a variety of products that can be used around the house, in the garden and as natural health and beauty products.

The natural scent of a favoured essential oil can waft through the house when used in cleaning products or scented candles, gently infusing the air. At the same time, that oil can help to rid the air of germs and bacteria and keep those coughs and sneezes at bay.

You can whip up an eiderdown from scraps of fabric to keep the winter chills at bay, or reuse those itchy, scratchy blankets that were once destined for the dog bed. Or make yourself a harvest apron to gather your bounty from the garden or from the pots of goodness growing inside when only a windowsill is available.

For minor ills and niggles, make a natural remedy your first port of call before diving into a packet of pills. For example, a buttermilk wrap is excellent for sunburn, or caraway tea is soothing for a jippy tum. Think about what you put onto your skin and opt for a more natural approach to cleanse it each day. Gentle almond and rosewater cleansing cream will give you a healthy glow, while the banish acne skin tonic is great to have on hand when dealing with a breakout.

The more you learn to use the things you have around the house in new ways, the more you will start to see potential in everything around you. Things that may have been thrown out can be used to nourish your plants or whiten the whites wash. Nature is, indeed, one of our greatest inspirations and it seems quite right for her to lead us through the simplest of daily tasks in a way that can make the place in which we live a little bit nicer than it was before.

house & HOME

OUR HOUSES, OUR HOMES, ARE A PLACE TO REST A WEARY HEAD, A PLACE THAT GIVES US COMFORT AND JOY, A PLACE TO RETREAT FROM THE WORLD AND TAKE JOY IN THE SIMPLE DAY-TO-DAY TASKS. MAKING A FEW CHANGES IN YOUR HOME CAN HAVE A PROFOUND DIFFERENCE ON YOUR DAILY CHORES. OVERHAULING YOUR CLEANING ROUTINE AND REPLACING HARSH CHEMICALS WITH GENTLE HOMEMADE VERSIONS CAN BE BENEFICIAL RATHER THAN HARMFUL TO YOUR WELLBEING. A FEW CRAFTY PROJECTS CAN LOOK GOOD IN YOUR HOME AND ARE GREAT FOR REUSING REDUNDANT BITS AND BOBS AS WELL AS BRIGHTENING THE MOST MUNDANE OF TASKS.

REDUCE YOUR NEED FOR PLASTIC BY MAKING A FEW SIMPLE ITEMS TO ADD TO YOUR KITCHEN. BY USING NATURAL MATERIALS, YOU CAN FEEL CONTENT THAT YOU ARE DOING YOUR BEST FOR YOUR OWN TINY PIECE OF THE WORLD. IT MAY SEEM LIKE IT HAS LITTLE IMPACT IN THE BIG SCHEME OF THINGS, BUT IMAGINE IF EVERYONE MADE A SIMILAR COMMITMENT. FROM TINY ACORNS MIGHTY OAKS INDEED GROW.

cleaning

BEFORE THE TIMES OF MANY, MANY SHELVES OF WONDER, of cleaners, sprays and magic solutions for keeping your home germ- and dirt-free, it was necessary for people to make their own, often from everyday products found around the house. Rather than using harsh chemicals to clean the home, many of these traditional recipes were far friendlier to the environment and got the job done just as well. Try your hand at cleaning without chemicals, whip up a batch of beeswax polish to nourish your wooden furniture, and learn a few old tricks to clean your house from top to bottom.

WOOD WIPES

Wooden furniture and dust are like the best of friends, separated for an instant but then back together as soon as you have turned around. These handy wipes help to keep the dust at bay and freshen up the room at the same time. Keep them in a sealed container to stop them drying out and use them as part of your war against dust.

INGREDIENTS
Roll of thick kitchen towel
500ml warm water
15ml white vinegar
15ml glycerine
5 drops essential oil

1. Using a sharp knife, cut the kitchen roll in half and remove the cardboard from the centre.

2. Place one cut half roll of paper into a container that has a lid; a large jam jar is ideal. Make sure the lid fits on tightly. Cut off any excess towel if necessary.

3. Place all the other ingredients in a jug and stir well until the glycerine has dissolved and the essential oils are well distributed within the mixture.

4. Quickly pour the mix into the container before the essential oils have had time to separate.

5. Secure the lid and leave the paper towel to soak up the liquid for 5–10 minutes. If there is liquid left at the bottom of the container or the towels seem overly wet, pour out any excess, then seal lid. The wipes are now ready to use. Carefully remove each one from the middle of the roll and wipe it over your furniture.

BEESWAX FURNITURE POLISH

A no-nonsense polish that will have any wooden pieces of furniture that have escaped the paintbrush looking their best. Use it every month or so to feed the wood and keep it in great condition. If you don't like the idea of using mineral oil it can be replaced with sunflower or olive oil, but the polish will have a shorter shelf life, as the plant oil will become rancid after a time.

MATERIALS
100g unscented mineral oil (baby oil)
20g beeswax
Few drops of essential oils

---

1. Place the mineral oil and beeswax into a glass jar, then place the jar in a heatproof bowl set over a pan of boiling water. Stir with a wooden skewer until all the beeswax has melted.

2. Remove from the heat and stir in the essential oils, ensuring they are well mixed into the oil and beeswax.

3. Place the lid on the jar and leave to one side until completely cool.

USING THE PRODUCT

Use this polish to revive your wooden furniture. Buff it on with a clean cotton cloth and leave it to dry for about 10 minutes. Reapply if necessary, or buff off any excess to remove tackiness from the polish, which will attract dust like a magnet. Obviously, before you embark on reviving a family heirloom worth a king's ransom by slathering this polish all over the place, perform a wee test somewhere discrete on the furniture.

SPRING CLEAN BUCKET

It's that time of year when the days begin to lighten and the dappled light of spring begins to shine into all those corners that have avoided anyone's gaze for several months. Dust is gathering and whispering behind the sofa and the moment has come to have a good old shift about and give everything a proper scrub. This project is an ideal companion as you move from room to room throughout your home, giving it a thorough once over.

MATERIALS
Large galvanised bucket
Length of sew/stick hook and loop tape,
the diameter of the bucket
2 good-sized napkins or tray cloths

---

1. Measure around the top edge of the bucket (fig 1), then divide this number in half. This gives you the length of each piece of tape. You may need to shorten each piece slightly depending on the placement of the bucket handle. Place the sticky piece of tape onto each side of the bucket, just under the rim (fig 2).

2. Using the sew-on piece of tape as a guide, measure along the top edge of your napkin or cloth. If you have been wise and crafty then your choice will be a perfect fit, but if it is a little too big, fold and stitch the fabric until you have the correct width. Stitch the tape along the top edge of your cloth.

3. Fasten the cloth onto the bucket and turn up the bottom edge of it to create a pocket (fig 3). Pin into place.

4. Stitch the sides and base of the pocket and do a row of stitching down the centre to create two smaller pockets in your cloth.

5. Repeat for the opposite side.

MEASURE
TOP EDGE

FIG 1

ATTACH
STICKY TAPE

FIG 2

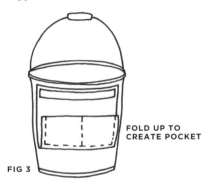

FOLD UP TO
CREATE POCKET

FIG 3

FOREST-FRESH FLOOR CLEANER

Ditching toxic cleaning chemicals from your home is always a good plan, for your well being and, indeed, your purse. This woodland-scented cleaner will make your floors sparkle and leave your home smelling of happy days strolling through a forest, crunching pine needles underfoot. Make it up in small batches to be added to your mop bucket or use neat along with some good old elbow grease on the more stubborn grime. Use ½ a cup in a bucket of hot water to mop your floor. This cleaner is okay on most non-carpeted flooring, but if in any doubt then try a little on an unseen area and leave it to dry before you mop the entire floor.

INGREDIENTS
250g washing soda
250ml white vinegar
100ml liquid castile soap
20–30 drops pine essential oil

1. Place the washing soda in a large bowl and put to one side.
2. Stir together the vinegar and soap until well combined.

3. Pour the liquid onto the washing soda and keep stirring until the soda has dissolved. You may need to add a splash of boiling water at this point.
4. Pour into a large plastic bottle, add the essential oils and shake well.

ALL-PURPOSE ORANGE CLEANER

Our grannies didn't have the vast array of cleaners for every eventuality to choose from when they did their weekly shop. They often made their own cleaning solutions from ingredients they already had in the home. With just a few items from your kitchen cupboard it is easy to make yourself an entire arsenal of cleaning products without any harsh or horrid-smelling chemicals. Below is provided a selection of recipes for products that will enable you to spruce up your home from top to bottom.

INGREDIENTS
Orange peel, enough to fill a jar
Bottle of white vinegar, to cover the
orange peel
Orange vinegar

---

1. Fill a large glass jar with orange peel and cover with white vinegar. Place the lid on and pop the jar in a pan of water and heat until the water reaches 65°C. Maintain at this temperature for 20 minutes, then take the jar out of the hot water and leave to cool.

2. Allow the jar sit for a few days, then strain the vinegar into a bottle. This product can be used on its own or in other recipes requiring vinegar.

CLEANING WITHOUT CHEMICALS

Once you are smitten with your all-purpose orange cleaner, why not add to your cupboard of homemade cleaning products by trying out a few of the easy recipes below. No fancy ingredients are needed, just a few straightforward items from the cupboard to make your home clean.

ANTI BACTERIAL SPRAY
100ml water
20ml orange vinegar
20 drops lavender oil

Mix together all the ingredients in a spray bottle. Spray this product onto surfaces and wipe down as and when needed.

ALL-PURPOSE CLEANER
30ml white vinegar
5g washing soda
10ml liquid castile soap
300ml hot water

Mix all the ingredients together in a squirt bottle and use for every eventuality.

FURNITURE WIPE
10ml homemade essential oil
(see page 126) – lemon works well
250ml water
100ml white vinegar

Combine all the ingredients in a large bottle. Squirt onto furniture and wipe with a dry cloth.

TOILET CLEANER
½ cup bicarbonate of soda
1 cup of white vinegar

Add the bicarbonate of soda to the toilet bowl and leave for at least an hour or overnight if possible. Pour in the vinegar and wait for the fizzing to stop then flush. Any tough stains may need a quick scrub with a toilet brush before you flush.

KITCHEN APPLIANCE CLEANER
50ml washing up liquid
1 heaped tablespoon cornflour
200ml warm water
200ml white vinegar

Mix the vinegar, water and washing up liquid together and then use a little of this liquid to mix a paste with the cornflour. Stir until smooth and then slowly add the rest of the liquid to the paste.

Pour into a spray bottle and shake before each use.

Spray onto your white goods and wipe clean with a damp cloth.

Discard any unused cleaner after a month.

CLEANING GLASS

There is nothing worse than when the sun shines and you realise just how grubby & streaky the glass is around the home, windows, mirrors and your beloved glassware can harbour dust and stains in the most tricky of places. Use any of the tips below to keep everything sparkling and clean.

FOR GREASE-MARKED WINDOWS OR MIRRORS, cut a lemon in half, then rub it over the area to be cleaned. Now spray with a 1:1 solution of water and vinegar and rub with a dry cloth. Then polish with a piece of crumpled newspaper (the ink will make the glass shine).

TO KEEP VASES OR DECANTERS SPARKLING, fill them with a few crushed eggshells and some vinegar or lemon juice. Leave the liquid in the glass for a couple of days, swirling the mix every now and again. Pour out the liquid and wash with hot soapy water, rinse, then dry.

GLASS VASES THAT HAVE BEEN STAINED with green algae can be cleaned with a mixture of black tea leaves and lemon juice. Place the lemon juice and tea leaves into the vase and rub with a bottle brush until the staining has gone. Pour the solution away and rinse well with hot water and soap.

RUB THE FRONT OF GLASS PICTURE FRAMES with half an onion or potato, then polish with a dry cloth.

LIMESCALE DEPOSITS ON GLASS can be removed by rubbing with a mixture of salt and vinegar… not just good on your fish and chips!

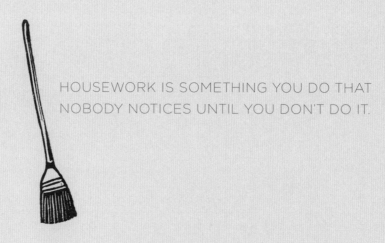

HOUSEWORK IS SOMETHING YOU DO THAT NOBODY NOTICES UNTIL YOU DON'T DO IT.

THE WONDER OF LEMONS

These zesty yellow fruit are well-known powerhouses for all things food and drink, but explore the list below and it will become clear that a large bowlful is an essential item to have close to hand in the kitchen for other reasons too. You may be surprised to find that you will be using them at least once a day and they will become your first port of call for many of the tasks in your natural home.

HANG SMALL CLOTH BAGS OF LEMON PEEL in doorways and around windowsills to keep various bugs and creepy crawlies away.

A FEW SLICES OF LEMON in a bowl of water placed into the microwave and heated for 3 minutes will loosen any baked-on food, which can then be easily wiped away.

MIX A SPOONFUL OF LEMON JUICE with a little olive oil and wipe down your leather shoes with a soft cloth for a nourishing shoe polish that will smell rather nice.

AMAZE THE CHILDREN with your secret agent skills and write a note in lemon juice then reveal it by heating it with a hair dryer.

WHITEN YOUR FINGERNAILS by soaking them in lemon juice for 5–10 minutes. Rinse with warm water, dry, and give them a quick buff.

MIX LEMON JUICE AND SALT and apply to copper pans with a soft cloth to remove any tarnish and get them looking clean and sparkling.

REMOVE LIMESCALE from draining boards and taps by rubbing them with lemon juice, leave for 5–10 minutes (or longer if needed) and then rinse with warm water.

PREVENT RICE FROM STICKING by adding the juice of ½ lemon to the pan whilst cooking.

FOR AN INSTANT AFTERNOON PICK-ME-UP, add some lemon oil to an aromatherapy burner and add a little spring in your step.

MAKE A GLASS OF LEMONADE by adding juice of a lemon and a heaped teaspoon of sugar into a glass, then add a splash of hot water to dissolve the sugar and top up with sparkling water. Add ice and a slice and relax with a good book.

laundry

BRING A LITTLE JOY INTO THE DAILY CHORE OF DOING THE
LAUNDRY. Luckily, washday in modern times is not the labour of love
that it used to be. With modern appliances it has become one of the
easier household tasks, but, sadly, that has a price and the gallons of
water and soapy chemicals that are churned out do their bit to pollute
the environment. So take a little of that saved time to do things the
old-fashioned way. Soak whites overnight or use a simple stain paste as
a first port of call before reaching for the harsher alternatives. Turn an
unwanted lampshade into a really nifty laundry basket, and freshen up
your fabrics before popping them in the wash.

HOUSEHOLD STAIN POWDER

Make up a batch of this simple two-ingredient powder to have on hand to pre-treat stains of all kinds before the item goes through the wash. No toxic chemicals, just simple kitchen-cupboard ingredients will help you rid fabrics of most stains. Store it in powdered form, then make it into a paste to use on stains and to brighten grubby grouting. Alternatively, mix it with vinegar or lemon juice to clean limescale deposits from taps, remove stubborn marks from saucepans and anything else that has developed an unsightly tinge.

INGREDIENTS
1 cup cooking salt
1 cup bicarbonate of soda

---

1. Place both the ingredients in a jar, put on the lid and give it a good shake to combine the salt and bicarbonate of soda.

As long as this is made in a 1:1 ratio the size doesn't matter, so make as little or as much as you need.

USING THIS PRODUCT

To use on clothing stains, make a small amount into a paste with some warm water, cover the stain with this paste, and leave it on overnight. Then wash the garment as usual. Remember – for precious clothing, try a small test patch somewhere out of sight, just to be on the safe side. For household jobs, mix the powder with either water or a mixture of vinegar and lemon juice.

STAINS

The best way to deal with a stain is speed. A soft cloth or kitchen roll is your first line of defence. Pressing it gently on the stained area can often remove most of the damage. If the stain won't budge, consider applying a patch or covering with decorative stitching. Below are some tried and tested stain removal methods, but always test a small unseen piece of fabric before embarking on a fully fledged stain battle, and, for an expensive item, call in the professionals.

BASIC STAIN-REMOVAL KIT

Kitchen roll

Soft cloth

Cotton wool – pads and buds

Wooden lolly sticks – for scrapping

Nail brush

Scrubbing brush

White vinegar

Lemon juice

Glycerine

Salt

Cornflour

Methylated spirit

GREASE AND OIL STAINS

Sprinkle the stain with cornflour and allow it to be absorbed by the flour, leaving it for at least 30 minutes. Brush off the flour, then wash the fabric in the washing machine using the hottest setting possible. (Alternatively, apply neat washing-up liquid to the stain and wash as normal.)

FRUIT JUICE OR COLA STAINS

Soak the stained fabric in milk for a least an hour, then wash as normal.

BALL-POINT PEN STAIN

Dab with a cotton bud dipped in lemon juice, then wash as normal.

MUD STAINS

Allow the mud to dry, then brush it off the fabric. Rub the stained area in cooled water from boiled potatoes, then wash as normal.

SWEAT MARKS

Apply a 1:1 solution of white vinegar and water and leave the fabric to soak for at least 30 minutes, overnight if possible. Wash as normal.

NAIL POLISH STAINS

Soak cotton wool in a non-oily nail polish remover and apply it to the stain. Wash as normal.

COFFEE STAINS

Rub with glycerine and rinse with warm water.

BLOOD STAINS

Soak the fabric in cold salty water, then use a biological washing powder to wash it.

GRASS STAINS

Don't use water as this fixes the stain. Dab it with a cotton pad soaked in methylated spirits, then wash with warm soapy water.

TOP TIPS FOR KEEPING YOUR WHITES WHITE

In the past, tradesmen specialising in the art of stain removal could be found peddling their wares from village to village. The solutions they offered took a more natural approach than we follow now. As is often the case, however, the old ways are still the most efficient and by using a few products such as vinegar or lemon juice, as well as good old sunshine, your whites wash will be looking its best in no time.

ALWAYS SEPARATE your whites from your coloured garments.

PRE-TREAT STAINS either with a splash of concentrated washing liquid or use the stain remover paste shown on page 27.

WASH LINEN AND COTTON in the hottest wash recommended for the fabric at least once every three washes (no matter what the miracle washing powder claims).

ADD LEMON, white vinegar or a small muslin bag of eggshells to your wash.

BOOST YOUR WASH with a couple of tablespoons of washing soda or borax.

HANG YOUR WASHING in bright sunlight to dry, which is especially effective if you have added lemon juice to the wash.

MAKE SURE YOUR WHITES are well rinsed — consider an extra rinse program if your washer has one.

WOOL DRYER BALLS

Although pegging your laundry out in the sunshine on a nice breezy day is the best way to dry your clothes, more often than not many folks don't have that option and have to rely on a tumble drier to get the job done. So to reduce the time your washing is in the drier, why not rustle up a few of these wonder wool balls? They help to soften your laundry so you can skip the fabric conditioner, they retain some heat and separate the washing, helping it to dry faster, plus cut down on static and so reduce the creases. All these things save time and money and are also good for the environment.

MATERIALS
100% wool roving
(a 100g hank should make 5 balls)
old pair of tights
large eye sewing needle

1. Begin by winding the wool around two fingers until you have started to form a tight bundle. Slip the wool from your fingers and keep winding the yarn around, moving the bundle in a clockwise motion so that you start to form a ball shape. Keep the tension tight.

2. When the ball is about 6cm across, cut the yarn and, using the needle, sew the wool through the centre of the ball and then back through to secure the end so that the ball doesn't unravel. Repeat for the other balls.

3. Cut off one of the legs of the tights and place the balls in one by one, securing each one with a knot.

4. Place in a pan of boiling water with a little washing powder added and boil for 20 minutes to make sure that the balls are thoroughly soaked with water.

5. Next pop the balls into the washing machine set to the hottest wash possible, preferably with some towels, as these will help the wool felt.

6. When the wash has finished, remove the tights and then pop the balls into the tumble drier to finish the felting process.

7. Leave them in the drier and, if you fancy a bit of scent, add a drop or two of essential oil to each of the balls, then use.

WIRE LAUNDRY BASKET

In a house full of family there is always a pile of dirty washing somewhere or other and those plastic laundry baskets, while practical, aren't very pleasing to look at. All too often large standard lampshades are left unwanted on the charity shop shelves, their pink satiny goodness no longer required. Having amassed quite a collection of these shades and only having one standard lamp to sit them on, it was decided that the time had come to bestow on them another job.

MATERIALS

Large standard lamp shade

Piece of chicken wire that's long enough to wrap
around the shade and cover the base

Sturdy pair of gloves

Fabric, yarn or suitable string for wrapping wire
edges and rope

Wire cutters

1m rope for handles – window sash cord is an
excellent choice

All purpose glue

Pliers

---

1. Prepare your lampshade by removing all the fabric, bits of glue and the inside whatnot that the lightbulb rests on. This should easily snap off with a few tugs.

2. Make a rough paper template for the base of the basket (which would have been the top of the lampshade) and place this onto your chicken wire.

3. Put on a sturdy pair of gloves and cut out the shape of the template in the chicken wire using your wire cutters. Make sure you leave at least 1.5cm extra to wrap wire around the base.

4. Lay the wire base onto your lampshade and attach by wrapping chicken wire tightly around the base. Keep the chicken wire taut as you work.

5. Snip off loose wires as close to the shade as possible.

6. To cover the rest of the shade, attach your wire to one side of the shade. Bend the edge of the wire loosely around the frame at the starting point of the lampshade to anchor it. Wrap the chicken wire around the entire shade until you return to your starting point. Don't try to fix the loose end of the chicken wire to the shade at this point otherwise you may get into a bit of a muddle.

ENJOY THE SIMPLE THINGS FOR
THEY WILL BRING YOU PEACE.

7. Once you have snipped away your excess wire, start to push and pull the wire to make it fit the shade snugly, keeping it taut as you work. There will be excess wire but you will remove this in the next step.

8. Once you are happy with your shape, snip off excess wire leaving yourself enough to wind the wire around the top, base and sides of the shade, snipping off any excess wire ends close to the edge of the basket.

9. Wrap the edges of the shade in fabric so that none of the pesky wires catch your clothes.

10. To add the handles to the basket, cut your piece of cord in half so that you have 2 50cm pieces.

11. Loop one of the pieces of rope under the top of the basket and fold it back on itself until the cut edge reaches about half way (approx. 25cm) and glue the 2 pieces together. Repeat on the other end of the string and make sure that the cut edges meet in the middle; this will form your handle. Wrap the length of the handle in string tying a knot at the end and secure it with a little glue.

12. Repeat the above step for the opposite handle

HANDY TIP
Smaller shades can also be covered in this way to create handy baskets for all sorts of nik naks, from balls of wool to fabric scraps, books or children's toys. Group together different sizes for some handy and lovely looking storage.

FABRIC FRESH SPRAY

There are many sprays on the shelves claiming to banish unwanted smells and odours from fabrics, which they do, but not without leaving an unnatural scent that brings on a fit of sneezing and wheezing – all in the name of freshening up musty fabrics around the home! Hanging your un-washables out in the fresh air is always the best line of defence against musty odours but, failing that, try mixing up a bottle of this product and squirting it onto your mustiest of linens.

INGREDIENTS

2 tsp bicarbonate of soda
30ml boiling water
50ml unscented fabric softener
60ml white vinegar
Few drops of essential oil
10ml vodka

1. In a large jug or container, dissolve the bicarbonate soda in the hot water and add the fabric softener.
2. Then slowly add in the vinegar, a little at a time, as the vinegar and bicarbonate of soda will react.

3. When all the fizzing and bubbling has stopped, mix the essential oils into the vodka, then add the mixture to the other ingredients. Pour into a spray bottle.
4. Shake and spritz when needed… (it's probably best to do a small patch test on any precious fabrics or brand new sofas first).

kitchen

IN MANY HOUSES, THE KITCHEN IS THE HEART AND SOUL OF THE
HOME. The place where food is prepared to feed a hungry hoard, where
most of the organising and planning for the week ahead is done, where
a crafty cuppa is taken in a spare 5 minutes and where most folks gather
to eat and talk about their day. Adding a few homemade touches to this
all-important room can make any time spent there a little more special.
Jars and bottles that are well cared for will last a lifetime. Begin also to
banish plastic from the house by using the beeswax wraps to cover food,
and make a few muslin produce bags to store the herbs
and spices that you dry on your willow hanger.

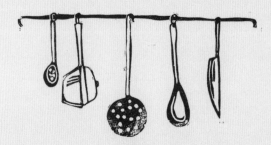

MARKET DAY TOTE BAG

There is always room for another bag in everyone's collection, no matter how many they have already, and here we have an ideal tote to take along on a shopping trip for a few bits and pieces. It is such a versatile little bag that you don't need to restrict its use to just shopping. It's also a great size for holding a knitting or crochet project and is easy to carry along with you here, there and everywhere.

MATERIALS
2 rectangles, W42 x L34cm (main fabric)
2 rectangles, W42 x L34cm (lining fabric)
Small square of fabric for patch pocket,
17 x 17cm (round off the 2 bottom corners)
Piece of jute webbing, W5 x L84cm
Bias binding, 50cm
Thread
Lacy embellishments or anything else
that takes your fancy

(Use 1cm seam allowances throughout.)

1. Hem the top edge of the patch pocket and press well. Unfold one edge of the bias binding and place right sides together on the edge of the pocket seam. Stitch along the unfolded crease on the edge of the binding, easing around the corners.

2. Fold the binding in half and turn it over the seam you have just sewn towards the wrong side of the pocket and press well. Place the pocket to one side.

3. Fold the piece of webbing in half and cut in two, measure 15cm from each end and mark with a pin. Fold length ways and stitch between the two marker pins. Press well.

4. Fold 2 rectangles in half and mark with a pin at the centre. Mark 5cm each side of this mark and 8cm down and pin piece of webbing each side of the centre pin to make the handle. Stitch a 5cm square at base of handle to attach it to the bag. Pin and place your pocket and edge stitch to secure to the front of the bag. Repeat on other side, omitting the pocket (fig 1).

5. Match and pin side and bottom seams and stitch. Press all seams open. Turn to right side and press again.

6. To make the base bring together the side and base seams at the bottom corner and flatten. Measure 8cm across keeping the 4cm mark inline

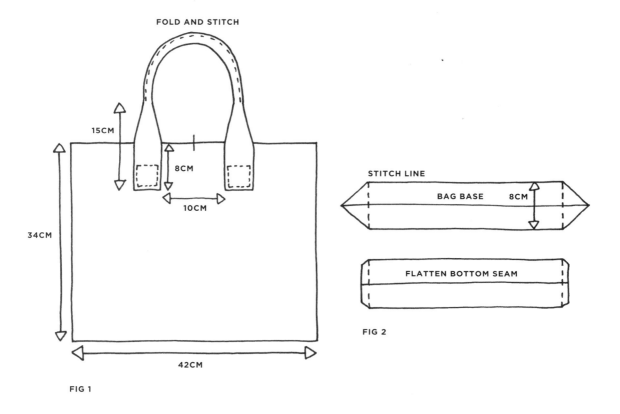

FOLD AND STITCH

15CM

8CM

10CM

34CM

42CM

FIG 1

STITCH LINE

BAG BASE   8CM

FIG 2

FLATTEN BOTTOM SEAM

with the seam and draw a line. Stitch across the line and trim (fig 2).

7. Repeat above step 2 steps for lining but leave an opening on the bottom seam to enable you to turn the bag to the right side. Press seams open.

8. Place lining over the main bag with right sides facing and ensure that the handles folded towards base and are enclosed between the main bag and lining.

9. Pin and stitch the lining to the main bag around the top edge. Press seam. Pull both sides of bag through the gap left in the bottom seam to the right side and stitch along the open seam.

10. Turn the lining into the bag and press the top seam flat. Embellish your bag by wrapping the handles, stitching on buttons, bows, and obviously lots of granny fake flowers.

REUSABLE BEESWAX FOOD WRAP

In these modern times trying to reduce plastic in our homes is a challenge, but this reusable wrap is a triumph in the banishment of clingfilm from the kitchen. Our grannies would have used a simple waxed paper or cloth wraps, but food wasn't expected to hang about for too long in a time when shopping everyday was the norm. In a few simple steps, you can make a reusable waterproof fabric in which to wrap all manner of foods and bowls of leftovers – and it looks pretty, too. Choose a patterned fabric or go for the natural look – just make sure it's a 100 per cent cotton and you can't go wrong.

Cotton fabric – a lightweight calico works well

Block of beeswax

Baking tray

Baking parchment

Old redundant cheese grater

1. Begin by washing and drying your fabric. Cut it into pieces that are sized and shaped to be suited to your needs. Try rectangles and squares for wrapping food, and circles to seal jars or bowls. For a neat edge and to prevent any fraying, use pinking shears, but it isn't necessary to if you don't have them.

2. Turn on your oven to its lowest setting and keep the door a little ajar. Beeswax melts between 62 and 64°c and will discolour above 85°c, so make sure your oven doesn't get too hot. Use an oven thermometer as a guide, if you have one.

3. Cover your baking tray with a piece of baking parchment, then place the fabric on top.

4. Grate beeswax over the fabric, trying to cover it evenly.

5. Pop your baking sheet into the oven and leave it for 5 minutes, keeping a close eye on it.

6. When the wax has melted, remove the tray and lift the fabric up to the light to check the coverage. The fabric should be almost opaque and you should be able to see the weave. Sprinkle more wax on any uncovered areas and pop it back in the oven if necessary.

7. When you are happy that the fabric is completely covered with wax, carefully remove the fabric from the baking tray and lay on a clean piece of baking parchment to cool. Once cool, the wrap is ready to use.

USING THIS PRODUCT

To use the wraps just use the heat of your hand to fold them around food, or cover bowls and jars. The stiffness of the fabric will keep them closed with no need for clips or ties. To keep the wraps clean and fresh, just hand wash them in warm tap water with a little soap and a bit of a scrub. Once the wraps are used and worn, make your self a new set and move the old ones onto the compost heap, or cut them into strips and use them to wrap kindling for fire-lighters.

JAR & BOTTLE CARE

Despite our throw-away-everything-plastic society many people are moving
back to using more traditional storage solutions such as glass and ceramics.
Well cared for preserving jars and pots can last a lifetime and will also look
far nicer on a kitchen shelf than a plastic container. They also have the added
bonus of not leaking back any chemicals into the contents that are stored
within them. Below are a few tips to keeping your glass in tip-top condition.

CLEAR UP YOUR CLUTTER FOR
IT WILL CLOUD YOUR MIND

REMOVE MUSTY SMELLS by adding a 1 teaspoon of bicarbonate of soda to each jar and filling with hot water. Place on the lid and shake well, leaving the contents to cool, preferably overnight. Rinse and then dry thoroughly and repeat if there is any lingering smell left behind.

WHEN USING KILNER OR MASON-STYLE JARS for preserving always use a new lid each time. If you are just using them to store dry goods then reusing the lid is fine

TO REMOVE OLD LABELS OR STICKY RESIDUE from jars you would like to reuse, fill with boiling water from the kettle, place the lid on tight and soak in boiling water for at least 20 minutes. The label should just slip right off, along with the glue.

IF YOU HAVE TROUBLESOME STICKY GLUE left behind, rub a little olive oil onto the problem area and let it soak in for at least 30 minutes. Wipe off with a piece of kitchen towel or scrape off the remaining glue with a blunt knife and wash in hot soapy water.

WILLOW WIRE HERB HANGER

Harvesting and hanging up bunches of herbs to dry is an age-old tradition that has been used for generations in order to preserve them to be used at a later date when they couldn't be found fresh. Somewhere well-ventilated and out of direct sunlight is all that's needed for drying. Why not make this simple willow hanger to keep them all together? As well as being an excellent addition to a kitchen or pantry, your herbs will be dry and ready to be stored away in no time. Willow can often be found growing on roadsides and in wild spaces, enabling you to harvest a withy or two. Alternatively, you can order some withies online and maybe plant a few in your garden.

MATERIALS
2–3 willow withies
Roll of garden/florist's wire
Wire cutters/secateurs

1. Bend a piece of willow into a circle and twist the ends over each other to secure it into a circle. Add the rest of the willow in the same way, twisting in the ends to make a neat circle.

2. Then, at four opposite points on the circle, wrap a piece of garden wire around the willow to keep it all neatly in place.

3. Using the secateurs, snip off any unruly ends that refuse to stay tucked in.

4. To form the hanging wires, cut off a long length of wire and wind one end around the willow in the space between the two pieces of wire used to secure the willow and leave the longer piece of wire facing upwards. Repeat until you have four wires.

5. Pull the tops of the wires together and twist several times to secure them.

6. Bend the top end of the wires down to form a hanging loop. Wrap another piece of wire tightly around the end of the loop to secure it.

7. To attach your herbs, tie them in bundles with a rubber band or a piece of string and use a butcher's hook or create an s-shaped piece of wire to hang them onto the willow.

8. Place your hanger in a well-ventilated area out of direct sunlight and leave the herbs to dry until they crumble when pressed between your fingers.

9. Remove the dried bunches from the hanger and store herbs in an airtight container.

HARVEST APRON

An extra pair of hands is always welcome when doing household chores or working in the garden, and this practical apron design originates from the 1940s, when ladies knew how to rock the multi-functional make-do-and-mend lifestyle. It doubles up as an everyday apron and, with a pull and a twist, a practical basket-type pouch can be created for carrying all sorts of everything. As the name implies, it is particularly useful when you are harvesting fruit and veg in the garden, leaving your hands free to pick the produce and pop it into the pouch, but it works just as well in the house, especially for collecting the randomly discarded items of clothing, toys and that odd sock that have all somehow found their way behind doors, down the back of sofas or abandoned in a trail across the bathroom floor.

MATERIALS
Lining or brown paper, to make a pattern
1.5m heavy cotton/linen or denim
2.5m 1cm-wide twill tape

(Use a 1cm seam allowance throughout.)

1. On your pattern paper draw a rectangle that is 45cm wide and 85cm long. Mark points A and B as shown in figure 1.

2. Draw a line from point A to B and then continue on to point C, curving the line at the corner to make the apron shape (fig 2). Cut out the pattern.

3. Fold your fabric in half lengthways, with right sides facing in, and selvedges together. Place your pattern onto the fold, and then draw around it onto your fabric. Cut out your fabric.

4. Cut out a piece of fabric measuring 12cm by the width of your fabric for the waistband/tie, and 2 strips measuring 1.2m x 5cm, cut on the bias. If you would like to add a small pocket onto the front of the apron, cut a rectangle measuring 20 x 25 cm. Hem all the sides, then stitch it onto the front of your apron in your desired position.

5. Prepare the bias strip by folding a 1cm hem at one end and stitch across to keep it in place.

6. Pin the bias strip right side to the wrong side apron piece (fig 3), unfinished bias edge to the waistband edge and stitch them into place. Make sure that the two neatened edges of bias meet at the bottom middle of the apron, leaving a space for to pull up the tape and form the pocket.

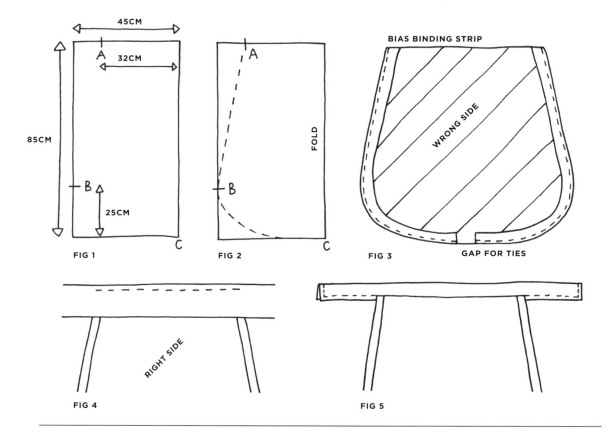

FIG 1

FIG 2

FIG 3

BIAS BINDING STRIP

WRONG SIDE

GAP FOR TIES

FIG 4

RIGHT SIDE

FIG 5

7. Press seam and then turn your stitched strip to the front of the apron to create a tape casing. Pin and stitch.

8. Thread a piece of twill tape through each casing and knot at the bottom. Measure 10cm down from the waist and secure the tape in the casing by stitching back and forth several times.

9. Pin your waistband/tie piece onto the apron front, matching centres and stitch together (fig 4). Fold the piece upwards and press seam downwards so that they lie flat.

10. Fold the waistband tie in half, right sides together and stitch down the short side and along the bottom of the tie until you reach the apron (fig 5). Repeat for the other side.

11. Snip any excess fabric from the corners and turn the tie to the right side. Repeat on the opposite side. Press the ties, making sure you have turned under the seam allowance on the part of the tie attached to the apron. Stitch along this opening and around the rest of the tie to give a neat finish.

12. Pop on your apron and go harvest.

COTTON PRODUCE BAGS

Make yourself some nifty cotton bags to use and reuse when you are buying your fruit and vegetables. Hang them from a hook or two in your kitchen to keep the produce fresh and close to hand. They make a great decorative addition to your kitchen, too. To store dry goods in your kitchen cupboards, or dishcloths, towels and any miscellaneous items that need organizing or hiding away, make up these bags in a thicker cotton or linen.

MATERIALS

50cm cotton muslin – will make approximately
4 bags, depending on the width of the fabric
Spray starch (optional)
Matching thread
String or thread

(Use 1cm seam allowances throughout.)

1. Fold the muslin fabric in half, with selvedges together, and iron out all creases. Because muslin can be quite tricky to work with, spray and iron dry several coats of spray starch. This will stiffen the fabric, allowing you to cut and stitch with ease.

2. For one bag, cut two pieces of fabric 25 x 30cm, then pin them together with right sides facing and stitch the two sides and the base leaving an opening on the side seam 3cm from the top edge (fig 1).

3. Snip to the stitch line on each side of the bag and then zig-zag or overlock stitch to neaten the seam and prevent the fabric fraying.

4. Press the seam open (fig 2) and then stitch around the edge of the opening on both sides of the bag.

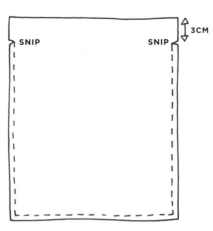

FIG 1

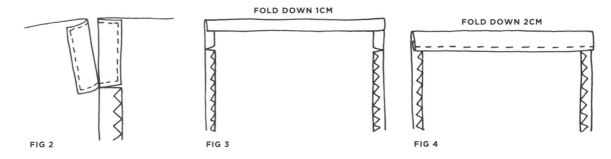

FIG 2   FIG 3   FOLD DOWN 1CM   FIG 4   FOLD DOWN 2CM

5. Fold the top edge on one side down by 1cm and press. Repeat on the other side (fig 3).

6. Now fold down each side by 2cm to make the casing for the string. The fold should be in line with the top of the neatened seam (fig 4).

7. Stitch along the bottom edge of the casing on each side.

8. Thread a piece of string through each of the casings and knot at either end, leaving 2–3cm loose.

9. Place your produce in the bag, pull it shut and hang it on a rather lovely hook.

around the house

THERE IS NOTHING LIKE A FEW CRAFTY MAKES SPRINKLED AROUND YOUR HOME to make it cosy and inviting, and feel like a place to relax and shake off the stresses of the day. Make rooms or projects quite individual as you try your hand at mixing your own paint, snuggle up under a bits and bobs eiderdown or turn a beautiful old wool blanket that does nothing but make you itch into a useable lap cover. Scent your home with natural fragrances that can help calm body and soul and have a few easy firelighters on standby for when the nights turn nippy and a roaring fire or woodburner is the only thing to get your toes toasting.

HOMEMADE PAINT

The idea of creating a paint from everyday ingredients found in your kitchen cupboards, which are non-toxic and safe to prepare, seems rather appealing. This paint is not suitable for all surfaces and will not cover emulsion or other paints, so don't make up huge pots to daub about here, there and everywhere, only to be disappointed by the results. Also, it will only keep for a day or two in the fridge. Bare wood, plaster, terracotta and clay are suitable surfaces, and it will need sealing with an appropriate wax finish. Have fun mixing up various shades and colour schemes and don't be afraid to experiment with all the ingredients until you get a finish you are happy with. Note down any blends of pigments you use, so you can recreate the colour in the future.

MATERIALS
2 tbsps artist powdered pigment
1 tbsp bicarbonate soda
6 tbsp hot water
4 tbsp quark (found in the cheese
section in supermarkets)
Beeswax polish, if you wish to seal the surface

1. To begin you will need to prepare your pigment the day before you intend to make the paint. Place your pigment into a small bowl and mix in just enough water to make a runny paste with the consistency of single cream. Leave the paste to stand overnight, then it is ready to add to your paint mix.

2. Dissolve the bicarbonate of soda in the measured hot water and leave to cool.

3. Place the quark into a large jam jar, then stir in the cooled bicarbonate of soda mixture. Finally, stir in your pigment paste until it is well combined with the quark binder.

4. Leave for an hour, then stir well again before using.

5. Leave each coat to dry for a few hours, and layer up the paint on the surface until you have a colour you are happy with. Leave for several days to fully dry and allow the paint to cure, then gently sand the surface and coat with a suitable wax sealant. The beeswax furniture polish on page 14 is a fine choice.

BITS & BOBS EIDERDOWN

Who could resist the old fashioned charm of a pretty, feather-filled eiderdown sat at the end of a bed? Eiderdowns in good condition are rather hard to come by nowadays and if you do find one, they usually come with a pretty hefty price tag. Along with the whole difficult-to-clean dilemma, they are often on display rather than in daily use. So make one of these splendid eiderdowns using those hoarded fabric scraps and snuggle up under a little bit of childhood nostalgia from a time when duvets were unheard of...

MATERIALS
100 fabric pieces, 17 x 20cm, for the pockets
2 rectangles, 1.5m x 22cm
2 rectangles, 1.1m x 22cm
Polyester wadding, filling fleece or feathers

---

1. Place two of the fabric pieces together with the right sides facing in and stitch a 1cm seam around 3 sides, leaving the fourth side open (fig 1).

2. Snip the excess fabric from the corners, then turn the pocket out and gently ease out the points.

3. Turn in a 1cm seam to the wrong side on the open edge and press.

4. Stuff your pocket with your chosen filling, being careful not to over fill it.

5. Hand- or machine-stitch the opening closed and repeat until you have 50 pockets. You may wish to give the pocket seams a gentle steam with the iron at this point.

6. Take the four fabric rectangles. Fold each in half along the longer length, with the right sides facing in. Pin and stitch a 1cm seam across both short sides and down the length to create a tube, leaving an opening about half-way along the long side to enable you to stuff the tube with filling (fig 2).

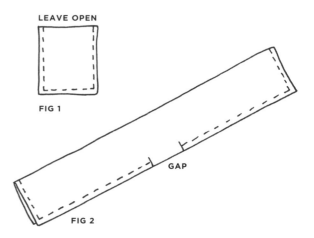

LEAVE OPEN

FIG 1

GAP

FIG 2

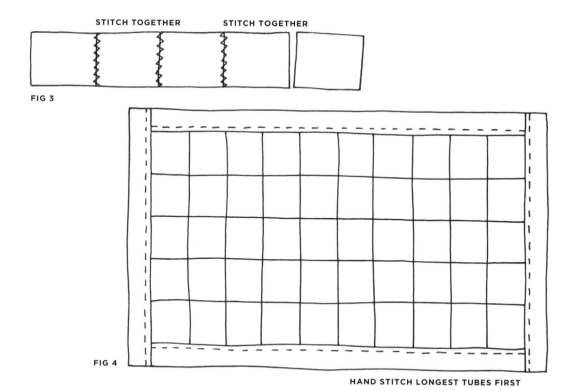

STITCH TOGETHER   **STITCH TOGETHER**

FIG 3

FIG 4   **HAND STITCH LONGEST TUBES FIRST**

7. Snip the excess fabric from the corners, then turn the right sides out. Gently ease out the corner points. Evenly stuff a tube with your filling. Don't over fill. Hand-stitch the gap closed. Repeat for the other three tubes.

8. When you are ready to construct your quilt, start stitching the pockets together in horizontal rows (fig 3). Either hand- or machine-sew the longer sides together until you have five pockets in a row. Repeat this until you have 10 rows of five pockets. If you choose to machine sew, use a zig-zag or similar and carefully butt two longer edges of the pockets together and stitch across the two edges, making sure each side is firmly attached.

9. Now stitch each row of five pockets together until you have 10 rows of pockets attached to each other.

10. Pin the two longest tubes down the sides of your quilt. Hand-stitch the edges together.

11. Lay the remaining two tubes across the top and the bottom of your quilt and repeat step 10 (fig 4).

12. Gently steam and plump the entire quilt.

CANDLE IN A JAR

Scented candles bring a sense of wellbeing and comfort to your home, especially on those dark, dank winter evenings when a gentle scent and a pleasant glow flickering in the corner of the room can lift the spirits no end. Making your own candles is really quite gratifying. Although it may seem an unnecessary task given the huge array of candles on the market, try it once and you will be hooked.

INGREDIENTS

Soy wax – enough for 2 jarfuls of wax
Glass jar with a lid
Candle wick that is slightly longer than
the height of your jar
30–40 drops essential oil
2 straws
2 elastic bands

---

1. Measure out 2 jarfuls of wax into a heatproof container.

2. Pour boiling water into your glass jar and place to one side. The water will warm the glass and help your candle to set evenly. Once the jar is warm, pour out the water and dry the jar carefully. Attach your wick to the centre bottom of the jar. A small glue dot or piece of double-sided tape will help to keep it in place.

3. Melt the wax, either on a double boiler or in the microwave. If using the latter, keep a careful eye on the wax and don't allow it to overheat.

4. Remove the wax from the heat and stir in the essential oils, then carefully pour it into your glass jar. Use the two straws secured together with elastic bands, clamped around the wick. This will help to keep it in place whilst the candle sets. Place the candle somewhere warm to set. Don't let impatience tempt you to pop it in the fridge as it needs to harden slowly to set evenly. If, once hardened, the top of the candle looks a little uneven or has sunk slightly, warm it with a hairdryer. This should even it out nicely.

5. When the candle it completely set, trim the wick and place the jar lid on to store it, or light it. On the candle's first burn, make sure it is alight for long enough to allow the entire top of the candle to melt. This will ensure the candle burns evenly on subsequent burning.

TOP TIPS FOR SAVING ENERGY

Just making a few small changes can make a big difference to your household energy consumption, which is important for both your purse and the planet. Often the flick of a switch is all it takes.

WHEN THE TEMPERATURE DROPS and winter creeps in, close all the curtains in your house to keep the heat in and the cold draughts out.

PLACE LIDS ON YOUR PANS when cooking and make sure that your pan fits the size of the cooker ring.

DON'T OVERFILL YOUR KETTLE, just boil what you need.

DRY YOUR CLOTHES in the sun rather than using the tumble dryer or adding extra humidity to the house.

TURNING THE HEATING THERMOSTAT down by 1 per cent can reduce your heating consumption by around 10 per cent.

WATCH YOUR WASHING – washing at 60° uses 30 per cent more energy than washing at 40°.

USE ENERGY SAVING LIGHT BULBS everywhere – they use approximately 80 per cent less energy than normal bulbs.

REMEMBER THE NATTY DRAUGHT excluders your granny used to make with her old tights? Stitch up a couple of tubes of fabric and stuff them with paper, sawdust, fabric scraps or polyester filling and cut down on those under-door draughts. Attach them to the bottom of the door with a couple of hooks so you don't have to move them every time you open the door.

SWITCH OFF STANDBY and unplug all those many appliances when not in use or fully charged.

LAYER UP – add a jumper or two instead of turning up the heating. Wearing a hat indoors can keep you snug. Thermal undies are also something not to be scoffed at, and invest in some wool socks to keep those toes toasty.

A PILE OF BLANKETS AND THROWS next to the sofa can be snuggled under when you are settled in for an epic movie marathon.

HAVE CURTAINS HANGING at all your external doors to keep heat loss and draughts at bay.

DON'T FORGET THE SIMPLE HOT WATER BOTTLE – great for warming beds, laps and cold fingers when they have seized up and refuse to work no more.

HOMEMADE SMUDGE STICKS

Cleanse and purify the stagnant corners your home with these easy to make smudge sticks. These tightly bound bundles of dried herbs have their roots in Native American ceremony, but they are also used in many cultures across the globe for purification of sacred spaces. Slowly burning the dried plants can help to shift stale air and bad energy from the darkest corners of a room.

MATERIALS

Selection of garden herbs – rosemary, thyme, lavender, lemon balm, mint, mugwort, pine, yarrow are good choices

Strong natural string – don't use anything containing manmade materials as it may release toxic fumes when burnt.

---

1. Harvest your herbs on a dry sunny day. Try to cut them in similar lengths about 10cm long. You will need between 15-20 stems depending on how leafy the plants are.

2. Arrange the plant stems into a bundle. Don't skimp on the herbs – you need a nice plump bundle.

3. Wrap the string tightly around the bundle at one end and then tie a knot to secure.

4. Turn the herbs in your hands and start to wind the string around the leaves. Once you have reached the top start to wind back down to the starting point.

5. When the bundle is neat and the plants are tightly bound together, wind the string around the base and tie off with a knot. If you wish, repeat step 4 again to make everything really firm.

6. Place the bundle in a warm dry place and leave until completely dry. This may take 2–3 weeks.

7. To use your smudge stick, hold the end bound with string and place the opposite end in the flame of a candle. When the bundle starts to crackle blow out any flames, as you want the end to smolder not blaze. Hold the smoldering end over a small bowl and then gently "smudge" the smoke around the room by gently waving the bundle into the darkest corners.

8. You may need to gently blow on the end of the stick to keep it smoldering. When you have finished just rest the lit end of the bundle over the bowl and it will go out ready for its next use.

FABRIC-BACKED BLANKETS

Itchy vintage wool blankets are often found for pence languishing in a forgotten box in many a charity shop. In the day of the duvet, very few of us use proper sheets and blankets on the bed, so these beauties are often chucked out for the ease of a modern solution. Adding pretty fabric backing to these blankets makes them great to throw across your lap on a chilly evening without suffering the dreaded wool itch.

MATERIALS
Old woollen blanket
1m soft cotton lawn or other
fabric of similar weight
5m bias binding
Matching thread

1. Wash and dry the blanket on a wool cycle. Give it a good iron, using steam if possible, to remove the creases and to flatten out the blanket.

2. Mark out a 1 x 1.4m rectangle and cut out the piece from your blanket. If your blanket has any holes or damage, repair at this stage with a patch or a bit of darning.

3. Place the blanket onto your cotton lawn with right sides facing out and smooth everything out. It may help at this stage to weigh down the fabric – old scale weights or tins are ideal for this.

4. Tack the blanket and cotton fabric together along the edges, smoothing as you go.

5. Trim the cotton to the size of the blanket.

6. Unfold the bias binding and pin down one of the edges of the fabric side. Stitch the bias to the blanket and fabric. You may wish to tack all the layers together at this point to make stitching easier. Repeat on the opposite side.

6. Fold the bias over the seam you have just stitched and pin to the blanket side. Hand stitch the bias down.

7. Repeat for the remaining two sides but leave a little extra tape at each end to fold over and neaten the edge. Now make yourself a cuppa, grab a good book and snuggle under your cosy warm blanket.

KINDLING WRAPS

There's nothing quite like the roar of a fire, a mug of tea and a bit of knitting to start your day, but getting the fire going – and quickly – can be a bit of a fine art. Making up some of these kindling wraps will ensure success every time, and they also look really great sat in your kindling bucket next to the fire. You can also do away with those smelly shop-bought firelighters that always end up crumbling all over the place in a hazardous mess. If you don't have an open fire to sit and toast your toes in front of, then make a few to start your barbecue or summer campfire.

MATERIALS
Finely chopped kindling wood
or small dry sticks
Dried herbs and grasses
Tissue or newspaper, cut
into 5cm wide strips
Raffia twine
Beeswax, paraffin wax or old candle stubs
Old paintbrush

1. Gather together 4–5 pieces of kindling or twigs and a bunch of dried herbs and grasses and wrap them with a strip of tissue or newspaper. Secure the wrap tightly with a piece of raffia.

2. Repeat step 1 until you have a pile of wraps.

3. Melt your wax in a jar in a warm oven.

4. Lay the wraps on a piece of baking parchment to protect your work surface. Using an old paintbrush, cover the paper in the melted wax.

5. Allow the wax to cool and harden, then use one or two of these wraps alongside your usual kindling to get your fire off to a roaring start.

THE WONDER OF EGGSHELLS

Once you have read the following information, you will never throw away your eggshells ever again. Because, amazingly, the eggshell is a powerhouse of healthy minerals that can help you to clean your home and nourish your garden. If you decide to feed eggshells to your animals or, indeed, ingest them yourself, use shells from organic chickens and make sure you wash and sterilise them within an inch of their lives.

EGGSHELLS ARE GREAT FOR CLEANING stubborn stains from glass, especially in those hard-to-reach nooks and crannies of fancy glass vases. Place some eggshells and lemon juice in the item to be cleaned and leave for several days, swirling now and again. Empty the vase, wash with hot soapy water and leave to dry.

TO MAKE YOUR WHITES SPARKLE, place eggshells in a small muslin bag secured tightly with a knot or string. Pop the bag into your whites wash and banish all signs of grey.

PREVENT SLUGS AND OTHER CREATURES from destroying your plants by sprinkling crushed eggshells around the plants.

FEED EGGSHELLS BACK TO YOUR CHICKENS — they much prefer them to the oyster-shell supplement. Mix the crushed shells in with their normal food.

POP A HANDFUL OF CRUSHED EGGSHELLS into a hole before you plant tomatoes to help prevent calcium deficiency problems, such as blossom -end rot.

POKE A PIN THROUGH THE BOTTOM OF THE SHELL and fill with compost, then add your seeds and a sprinkle of water to use them for growing seedlings.

ADD EGGSHELLS TO THE COMPOST HEAP to boost the calcium content.

SOAK EGGSHELLS IN APPLE CIDER VINEGAR for a couple of days, then use the mixture on bites or itchy skin. This will last for a month in the fridge.

MAKE AN INDOOR PLANT FOOD by half-filling an old jar with eggshells, then topping up with water. Leave for a few days, then water your plants with the mixture.

MAKE YOUR OWN CALCIUM SUPPLEMENT by baking eggshells in the oven at 180°C for about 10 minutes. Leave them to cool, then grind them to a fine powder. Add a spoonful of the powder to soups or smoothies once a day. Can be stored for a month.

FILL HALF AN EGGSHELL with beeswax and a wick plus a few drops of your favourite fragrance to make candles.

HOMEMADE REED DIFFUSER

Filling your home with natural fragrance does wonders for your heart and mind and makes your home just a little more cosy and comfy. These rather natty reed diffusers don't need an open flame or a fancy heated stone to release their aroma around the place, just a nice looking bottle, some wooden skewers and a blend of your favourite oils. Try to stick with natural essential oils as you will benefit from their therapeutic qualities as well.

MATERIALS
5–6 thin bamboo skewers,
with the sharp ends removed
50ml surgical spirit
50ml light oil (grapeseed, sweet almond,
light olive oil)
Narrow-necked vase or bottle
30–40 drops of your favourite essential
oil or blend

1. Begin by removing the pointed ends of the wooden skewers with a pair of sharp scissors and pop them to one side.

2. Place the surgical spirit and oil into the vase or bottle and swirl around well to mix the two.

3. Add the essential oils and swirl again.

4. Place as many skewers into the bottle as will fit, but don't over fill as the skewers will need space to release the fragrance.

5. Leave for several hours, swirling the liquid every now and again, then flip the skewers so that the dry ends are in the liquid.

6. You can repeat the previous two steps until the wood is saturated with the oil and then flip the skewers every couple of days to keep the fragrance wafting.

HANDY TIP
The volume of oil and surgical spirit can be increased according to the size of your vase or bottle, but keep the ratio 1:1. Essential oils can be added at 10 drops to 100ml of base oil, but do feel free to experiment with this volume if you would like a stronger or weaker fragrance.

in the GARDEN

THE GARDEN IS BECOMING AN EXTENSION OF OUR HANDMADE

HOMES, EVER MORE POPULAR AS FOLKS ARE TURNING BACK

TIME AND CONCENTRATING THEIR EFFORTS ON CREATING A MORE

PRODUCTIVE GARDEN. DOING WITHOUT CHEMICAL PESTICIDES

AND FERTILSERS, WE ARE RETURNING TO THE OLD TRIED-AND-

TRUSTED WAYS SUCH AS KEEPING A FEW HENS OR ENCOURAGING

THE BIRDS AND BEES AND OTHER GARDEN ALLIES TO CREATE AN

ECOSYSTEM THAT CAN SORT ITSELF OUT WHEN THE PESTS MOVE

IN. THERE ARE ALSO SUGGESTIONS HERE FOR THOSE WHO DONT

HAVE OUTSIDE SPACE – IT'S AMAZING WHAT CAN BE GROWN ON

A SUNNY WINDOWSILL OR TWO.

HERBS

Below are my favourite top five herbs to grow and use. Ideal for growing on a windowsill if your garden space is lacking, these herbs will often come to use and not just for cooking.

LAVENDER

The shining light of the herb garden, fragrant and medicinal lavender is essential to the backbone of any plot. It needs a sunny position and a light sandy soil in order to thrive. Once harvested, use in linen sachets, home scents, soaps, and make lavender sugar, salt or syrup to use in the kitchen. In oil form, it is great for stings, minor scrapes and cuts and is second to none on burns.

LEMON BALM

This bushy perennial plant grows a little too easily in the garden so will need a firm hand to keep it in check. It has a fresh lemon smell with a minty undertone. Its small white flowers are rather attractive to bees. A favourite in the kitchen, lemon balm is especially good with fish and poultry and makes an excellent pesto. Throw a few leaves into hot water and you have a refreshing and uplifting tea that is perfect at around the 3pm slump time. Lemon balm can be used to make an excellent antiviral balm (see page 110), which helps to fight off the dreaded cold sore and can be used as a mosquito repellant.

PARSLEY

Easy and simple to grow, parsley is often banished to the humble role of garnish, but it has a wonder of uses and a bunch is dense with vitamin C. It is packed full of antioxidants and a small amount added to your daily diet can help you reap some of its green goodness. It seems to be everyone's favourite garnish.

Use it in salads, in pesto or combined with fennel to make an after-dinner tea to aid digestion. Chewing a sprig or two can neutralize and freshen the breath.

PEPPERMINT

Growing it in a pot is the easiest way to control this wayward herb, which can be used in a multitude of ways. It makes a cooling mint tea that is good for the digestion and calms an upset stomach. Used in the home, peppermint is an excellent pest repellant (to use for mice, rats, ants and other unwelcome bugs). Spraying an infusion of peppermint leaves in rainwater can be used to kill aphids on your plants. If your animals are unlucky enough to pick up a tick, a drop of peppermint oil onto the tick will help release its jaws so it can be easily removed.

ROSEMARY

This is a hardy evergreen shrub with a strong woody scent. Rosemary can be infused in an oil, which can be used on irritated skin or rubbed onto the temples to relieve a headache. It also makes an excellent hair rinse and can help with dandruff. Taken as a tea, rosemary can aid a speedy recovery from illness. In the kitchen, blended with salt it makes a great seasoning for all kinds of dishes and is especially good with chicken and as a topping on a homemade loaf of bread. Plant it next to your carrots to keep the flies away.

NOURISH YOUR SOIL NATURALLY

To grow healthy strong plants in your garden it all starts with the soil. Dry, dusty ground lacking in nutrients isn't the ideal place to grow anything but weeds. Ditch the chemicals that promise a bumper crop at any cost and go for a more natural approach. Add any of the natural fertilisers below and your garden will flourish, the fruit and vegetables will be bursting with goodness and there will be a warm glow in your heart that you let those helpful little critters live another day.

EGGSHELLS – sprinkle into a planting hole to provide young plants with extra calcium, then throughout the rest of the growing season sprinkle onto the ground both to help prevent slugs from devouring your prized lettuce and to add nutrients to the soil.

WOODASH – used sparingly, this adds potassium and magnesium to the soil, which are both nutrients vital for healthy growth. It can also help neutralise overly acidic soil. Only use ash from hardwoods and non-chemically treated wood.

COFFEE GROUNDS – sprinkle onto your soil to add some acidity, provide calcium, magnesium and nitrogen and deter those pesky snails and slugs.

VEG PEELINGS – fill a trench with these and cover with a layer of soil before planting your seeds. The peelings slowly break down, supplying the growing plants with a few extra nutrients to help them on their way.

HOMEMADE COMPOST – the workhorse of any healthy garden. Take care to make a good compost and your garden will shine.

SEAWEED – gather a little on your next jolly to the seaside and dry it, crush to a powder and then sprinkle around your plants.

EPSOM SALTS – tomatoes, roses and peppers can't get enough of this stuff, and sprinkled onto the lawn and then watered in it can give it a green boost and perk it up no end.

GARDEN FRIENDS

After growing healthy, strong plants that can ward off attacks from pests and diseases, the next step to making your garden a healthier, chemical-free zone is to entice and encourage a whole host of friendly critters to dwell, often unseen, amongst the flora and fauna. There will need to be a few pests hanging about to provide substance for your friendly critters, but this is an important part of establishing a natural balance in the garden and keeping soil, plants and growing conditions the best they can be.

PROVIDE DARK, DANK SHELTERS for various insects, such as a log pile, a few large stones, hollow stems of plants or a pile of undisturbed leaves in an unused corner. They will also need a food source of nectar and pollen, so add a few plants to your garden with that in mind.

FROGS AND TOADS are invaluable for keeping pests in check and like a damp corner or small pond.

HEDGEHOGS ARE FABULOUS garden visitors and munch their way through a slug population in no time. They like a dry, undisturbed corner, preferably in a hedgerow or a nice dry pile of leaves, and a nearby water source.

BIRDS NEED SOMEWHERE TO NEST and thick-covered walls of dense climbing plants, large shrubs or trees. They will eat a variety of insects, aphids, bugs and spiders. Birds such as sparrows and finches are rather keen on a weed seed or two.

ESTABLISHING A NATURAL SYSTEM of pest control in your garden can take time. Below are listed a few non-chemical ways to wage war on unwanted visitors.

EARWIGS CAN BE DEVILS for munching on your newly germinated seedlings. An upturned plant pot filled with dry grass or straw can be used to catch earwigs and remove them from your precious vegetable garden.

HALF AN UPTURNED ORANGE or grapefruit can be an irresistible hiding place for slugs, which can then be caught and disposed of.

COVER YOUR BERRIES AND CHERRIES with old net curtains to prevent the birds from eating them all before you get a look in. It also encourages them to eat the things you don't want to keep in the garden.

COPPER TAPE around pots can deter slugs and snails.

WINDOWSILL SALAD

Growing your own produce is a most fulfilling activity but, sadly, not everyone has access to space that is suitable. Fear not, for a humble windowsill can be turned into a wealth of health and nutrients buy growing a few simple plants that can supply you with fresh, succulent leaves – snip off a handful or two whenever you fancy. All you need to get you going is some good-quality compost, a variety of salad seeds and a suitable container.

MATERIALS

Pebbles or a few pieces of broken crockery
A rectangular plant pot with a tray
Compost – potting compost is fine
Seeds – varieties of the cut-and-come-again type are ideal

---

1. Place your pebbles or broken crockery into the base of your pot and then cover with the compost.
2. Moisten the soil and allow it to settle, topping it up it as necessary.
3. Sow the seeds in lines along the length of the pot. Cover them lightly with compost.
4. Place on a sunny windowsill and water regularly to keep the soil moist.

5. Once germinated, you may need to thin the seedlings. If using the cut-and-come-again varieties of salad leaves, you can grow them much closer together as you will be cutting the young leaves. It is possible to get at least four cuts from some plants, but when the growth has slowed, remove the plants, refresh the compost and sow more seeds. This way, you will have a constant supply of delicious, healthy salad.

PLANT TONIC

Give your young plants and seedlings a head start by feeding them this simple tonic. Weeds are great at extracting various nutrients and minerals from the soil, so unlock their hidden potential and put them to work in this healthy garden brew. Add various other plants that may be heading for the compost heap into the mix to give your tonic an extra bit of zing.

MATERIALS

Old pair of tights or a large net bag

Selection of garden plants and weeds – nettles,

comfrey, dandelion, dock, parsley, mint, yarrow,

plaintain, rosemary, thyme, mint,

fennel and so on

Large bucket

---

1. Stuff a leg of the tights or the net bag with your garden plants. Pack them in as tightly as possible and then knot the top to secure them inside.

2. Place this in a large bucket of water, cover and leave for at least three weeks, stirring it from time to time. It will smell, so be brave. When the tonic is ready, use it at a ratio of 3:1 water to tonic.

COMPOST BOOSTER

Making good compost is indeed an art. If a heap of plant matter is to be turned into beautiful, rich growing soil, then it needs to be treated with care. A balance of brown and green stuff is essential. Too much green and your heap will turn to slime and stink. Too much brown and you could be waiting for a long time for anything to happen.

If you have pondered the contents of your heap and feel that it is pretty balanced but seems to be lagging, adding a bit of a boost can be just the thing to get things going. I suggest adding any of the following to really heat up your heap: old beer, coffee grounds, comfrey, a sprinkle of grass clippings, chicken manure and urine. Or use the recipe to brew up some booster and really get things rotting.

MATERIALS
200ml rainwater
100ml glycerin
Large jar with lid
2 tbsp mixed dried herbs – nettle, chamomile,
dandelion, yarrow and so on
Few pieces of oak bark – gather from logs or
the base of the tree, don't peel the fresh bark

---

1. Mix the water and glycerin together in a large jar, put the lid on and shake well. Add in your dried herbs and oak bark, give it another good shake and leave it in a warm, sunny place for at least a week.

2. When you are ready to add the mix to your compost heap, use a pole to make 5–6 holes in your heap, then pour about 50ml of the booster into each hole. Cover over the hole with soil and press down firmly. Repeat every six weeks or so.

DECAL PLANT MARKERS

Plant markers are a most practical and sensible thing to pop into a plant pot; they will remind you of all the information you need to remember about each particular plant. They are an essential piece of gardening paraphernalia, but that doesn't mean that they have to be dull. With a touch of a floral decal these handy-sized markers have plenty of room to write everything you need, and then can be cleaned off and used again.

MATERIALS
1 packet of air-dry clay
Marker template (see below)
Ruler
Sharp knife
Sheet of medium sandpaper
Sheet of small floral decals
Clear varnish
Chinagraph pencil

---

1. Roll out the clay into a flat rectangle that's slightly larger than A4.

2. Starting at one corner, lay the marker template onto the clay and carefully cut around it. Repeat until you have eight markers.

3. Smooth any lumpy edges with the ruler and leave to dry.

4. Once the markers are completely dry, smooth the edges with sandpaper.

5. Prepare your decals according to the instructions on the packet. Carefully apply one to the top of each marker.

6. When dry, give each marker at least three coats of varnish.

7. Use a Chinagraph pencil to write on the markers as needed.

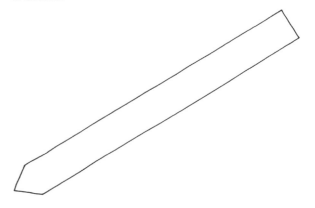

WHAT'S IN SEASON

To get the very best from your produce, try to buy seasonally. Sadly, that does mean no strawberries in winter, but you will enjoy the sweet goodness of the superior fruit when its time comes around and you take that first bite of the season. Below is a brief list of produce, explaining when each item is at its best in each season. Bear in mind that there may be regional differences. This list provides just a start to get you on the right track.

SPRING

VEGETABLES – Broccoli, spring cabbage, spinach, cauliflowers, leeks, radishes, salad leaves, kale, spring onions, watercress, early broad beans.

FRUIT – Rhubarb.

SUMMER

VEGETABLES – Lettuce and salad leaves, rocket, beans, beetroot, broccoli, carrots, courgettes, cucumber, fennel, peas, radish, tomatoes, garlic, watercress, kale.

FRUIT – Strawberries, blueberries, currants, elderflowers, raspberries, melons, citrus, gooseberries, loganberries, tayberries.

AUTUMN

VEGETABLES – Celeriac, carrots, mushrooms, kale, leeks, marrow, courgettes, pumpkin, squash, chard, sweetcorn, tomatoes, beetroot, salad leaves, beans.

FRUIT – Apples, damsons, plums, pears, quince, sloes, elderberries, blackberries.

WINTER

VEGETABLES – Brussels sprouts, cabbage, cauliflower, celeriac, kale, leeks, parsnips, potatoes, red cabbage, swede.

FRUIT – Apples, pears.

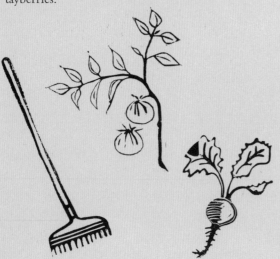

WILD FOOD

Mention 'food for free' and many an ear will prick up or head turn. In these days of modern convenience stores, growing or gathering our own food is something that is often passed by. If you look closely at your local pathways, quiet roadsides or wild spaces you can find all manner of fresh ingredients to make a tasty meal. Imagine wandering a country lane, basket in hand, rather than the trials of navigating busy supermarket shelves, feeling uninspired.

Many of us associate this gathering tradition with autumn when the hedgerows are bursting with blackberries, rosehips and the odd hazelnut or two, but there are many different edible plants that are on offer for most of the year. Opposite is a basic seasonal list to get you started, but always check what you are picking with a more in-depth guide book. Be wary of plants that you are unsure of and never just experiment with an odd leaf or berry. Also, don't forget to check if it is okay for you to gather produce from a park or wild space, especially if you are not sure whether it is public or privately owned. And be thoughtful. Don't strip an entire area – leave some for the wildlife and other gatherers.

SPRING
Chickweed, comfrey, young borage leaves, crab apple blossom, dandelion leaves, wild garlic, nettles, hawthorn blossom, mushrooms.

SUMMER
Angelica, borage (small leaves and flowers), chamomile, chives, comfrey, fat hen, elderflowers, lady's mantle, rose petals, sorrel, marigold, mint, tansy, wild strawberries.

AUTUMN
Blackberries, cobnuts, crab apples, elderberries, medlar, rowan, sloes, damsons, walnuts, hazelnuts, hawthorn berries, rosehips.

WINTER
Nettles, dandelion, rosehips, ground elder, alexanders, sweet cicely, chickweed.

HOW TO HELP BEES

Spring is a really tricky time for honeybees and indeed other pollinators in the garden. After eating all the stores over the winter the bees will need to replenish them quickly to help increase the number needed for a successful colony and a bumper honey crop. Below are a few things you can do to help bees in the springtime.

EMBRACE THE WEEDS – try to keep a small patch of your garden untouched and full of all those early and late-season food sources. White clover and dandelions are especially valuable at this time

CHUCK OUT THE CHEMICALS – all of them, even those pretending to be a-okay, they are all harmful to bees and will do them no good at all.

THINK ABOUT PLANTING COLOUR – honeybees can't see the colour red so ditch these plants from the top of your list and go for whites, yellows, blues and purples and plant in groups where a mass of flowers can be seen from the bees' flight path.

GIVE THE BEES A DRINK – place a large stone or pebble in your birdbath or a shallow bowl in the garden. Bees need water to drink but so that they don't drown they also need somewhere safe to land.

BUY HONEY FROM A LOCAL OR REGIONAL BEEKEEPER – it's more likely to be raw so much healthier for you. It will cost you more but you will be supporting your local bee population. Never leave unwashed honey jars outside; honey can be contaminated with bee disease spores, which could wipe out many colonies.

TEND, GROW & GATHER . . .
FILL YOUR DAYS WITH JOY.

THE WONDER OF HONEY

This golden liquid made from the nectar collected by hard working honeybees is truly a superfood and has a place on the shelf of everyone's cupboard. Honey can help heal the body both inside and out and is moisturising, antibacterial, anti-inflammatory and tastes rather good. Don't forget that not all honey comes from a trustworthy source and can be mixed with sugar syrup and other fillers. Buy local from a beekeeper if possible and then by region or country. Often the worst offenders in honey adulteration are those cheaper imported versions, so read the labels.

CLEAN CUTS AND SCRAPES – honey is a natural antiseptic. It has many antimicrobial properties and will help keep infection at bay.

DAB IT ON STUBBORN ACNE – honey will help fight any infection and aid the healing process.

USE IT AS A LIP BALM on dry chapped lips. Try not to lick it all off before it has done its job!

COVER BURNS WITH A THIN LAYER of honey and cover lightly with a damp cloth.

MASSAGE INTO THE ENDS OF YOUR HAIR for a nourishing treatment, leave for 20 minutes and then shampoo and condition as normal.

SOOTHE AN UPSET STOMACH by sipping a tea made from honey and ginger.

A TEASPOON OF HONEY EATEN RAW can help relieve the symptoms of indigestion and acid reflux.

EAT HONEY TO CURE THE HANGOVER from hell after a night of indulgence.

IF YOU SUFFER FROM HAY FEVER try to purchase local honey to eat on a daily basis, as it can help you build immunity to the allergens in your area.

BIRD-FEEDER BOTTLE

As summer draws to an end and the nights begin to darken, it's time to have a few thoughts for our feathered friends. As food grows scarce, we can provide them with some tasty nibbles to get them through the chilly days ahead. These repurposed plastic bottles make great bird feeders and look rather fine and dandy hanging in your garden.

MATERIALS

Small plastic drink bottle with screw-top lid
Piece of garden wire
Strong glue
Plastic plant pot saucer
White/coloured spray paint
Decal or sticker, if you fancy
Spray varnish
Birdseed
Wire cutters
Scissors

1. Wash and dry your bottle, removing all the labels.

2. Unscrew the lid and pierce a hole in it that is large enough to feed your garden wire through.

3. Make a hanging loop with a piece of garden wire and feed it through the lid, twisting the ends to secure the loop on the underside of the lid. Screw the lid back onto the bottle.

4. At the base of the bottle, cut out a small arch shape on the side of the bottle, trimming carefully to avoid leaving any sharp edges.

5. Cover the base of the bottle in a strong glue that's suitable for plastic. Attach it to the centre of the plant saucer. Leave to dry.

6. When the glue is completely dry, spray the bottle with several coats of your chosen colour. Add your sticker or decal and give it a final coat of spray varnish to protect it from the weather. Fill with birdseed and hang outside.

KEEPING CHICKENS

There is never a dull day when you have a few chickens around the place, from the proud squawks after laying an egg to the funny looking scratch dance as they forage for food in the undergrowth. Chickens are a low-maintenance pet that, in return for an enjoyable free-range lifestyle, will reward you with tasty eggs for most of the year. Let loose on the exhausted vegetable garden in autumn they will scratch the surface, breaking up the soil, eating any pests and leftover plant matter, depositing a little fertilizer as they go. Below are a few tips that will help your chicken ownership go without a hitch

HANDLE YOUR HENS REGULARLY – your chance to have a hug and check their wellbeing at the same time.

USE SAND ON THE FLOOR OF THE CHICKEN HOUSE – it makes cleaning a breeze.

A FEW DROPS OF GRAPEFRUIT SEED OIL or apple cider vinegar in the drinking can do wonders for the general health of the hens.

USE CRUSHED-UP EGGSHELLS IN THEIR FOOD, rather than buying expensive oyster grit – the hens much prefer it.

FEED THEM A WARM MASH in winter of oats and bran mixed with warm water – it keeps them toasty warm on the inside.

DURING THE MOLT, when your hens are looking a little worse for wear, feed them some scrambled eggs (cooked eggs won't lead to a raw egg-eating habit). They are a great source of protein and jam packed with other goodness that your hen needs.

A HANDFUL OF PUMPKIN SEEDS every day could help prevent worms and, if not, the chicken love foraging for them.

natural HEALTH

USING NATURAL INGREDIENTS AND REMEDIES TO HELP EASE

OR WARD OFF ILLNESSES AND AILMENTS HAS BEEN A TRADITION

FOR CENTURIES. NOWADAYS, PEOPLE ARE NOT SO KEEN TO REACH

FOR A PACKET OF PILLS OR AN OVER-THE-COUNTER MEDICINE,

AND ARE LOOKING FOR ALTERNATIVES. MORE AND MORE OF

US ARE RETURNING TO THIS AGE-OLD KNOWLEDGE AND ARE

REALISING THE GREAT HEALING POWER OF HERBS, PLANTS AND

EVERYDAY CUPBOARD ESSENTIALS. HONEY AND VINEGAR ARE

STALWARTS OF A HEALTHY LIFESTYLE AND CAN BE USED FOR

VARIOUS ACHES AND PAINS. NATURAL CURES FOR SUNBURN,

A JIPPY TUMMY OR THE DREADED COLDSORE CAN ALL BE TRIED

BEFORE REACHING FOR THE CHEMICAL-LADEN ALTERNATIVES

THAT MORE OFTEN THAN NOT JUST DON'T WORK.

VITAL VINEGAR

Such a humble underrated liquid discovered by chance over 10,000 years ago after a cask of wine had turned sour. Vinegar has been used throughout history to preserve all manner of things – as a drink and a digestif, an early remedy for various ills, a great all-round household cleaner, for washing fruit and vegetables, cleaning wounds, killing lice and preventing scurvy, to name but a few. Below are a few ideas of how we can use this wonderful liquid in our modern-day lives.

GIVE BRASS, COPPER OR PEWTER A SHINE to be proud of by mixing 1 teaspoon salt with 1 cup of white vinegar and stirring in flour until it becomes a paste. Apply to the metal and leave for 20 minutes, then rinse it off and polish with a soft cloth until dry and shiny.

SOFTEN WOOL OR COTTON BLANKETS by adding a cup of vinegar to the rinse cycle of your wash.

STOP FRUIT FLIES IN THEIR TRACKS by setting a trap in your kitchen. Add 2 tablespoons sugar and 2 tablespoons of vinegar to a cup of water and leave on the side in an open jar to lure the flies in.

CLEAN OUT YOUR KETTLE and remove limescale build-up by filling it with a bottle of vinegar and leaving to stand for 30 minutes. This works for your iron and coffee-maker too.

REMOVE RINGS left from cups and glasses on wood by rubbing with 1:1 mix of vinegar and olive oil.

REMOVE LESS THAN LOVELY smells from your fridge by placing an open jar of vinegar on a shelf inside.

KILL PESKY WEEDS in the garden by spraying them with vinegar. Repeat on any new growth until the plants are dead.

ADD A 1:1 MIX OF SUGAR AND VINEGAR to your flower water to give them a few days of extra life before they start to droop.

IF YOUR DOG IS A BIT NIFFY, wash him down with water and then mix 1 cup of vinegar in a bucket of water and wash him with the solution. Leave his coat to dry and the niff will be gone.

AT THE FIRST SIGN OF A COLD drink 1 teaspoon honey, 1 teaspoon apple cider vinegar and 1 teaspoon lemon in hot water to help fight the dreaded sniffling nose and sore throat.

HOMEMADE APPLE CIDER VINEGAR

The wonders of using vinegar medicinally date back to when these sorts of things were first recorded. As well as making a rather tasty salad dressing, cider vinegar is a great natural antiseptic that can be used safely about the home in your cleaning routine. A teaspoon of honey with a splash of vinegar is said to do wonders for your health. Chickens and other critters can also benefit from a spoon or two in their drinking water. Ideally a juicer is used to get as much liquid from the apples as possible, but if you don't have access to one of these gadgets, a blender and a mesh food-straining bag will serve you just as well. Do search out proper cider yeast and don't be tempted to chuck in a bit of baker's yeast instead, for the results could be dire.

INGREDIENTS
2kg apples, preferably organic
Large jar
Cider yeast
Juicer, blender and straining bag or apple press
Piece of muslin
Elastic band

1. Begin by juicing your apples. If you are going down the blender route, chop the apples and blend them to a fine pulp. Leave in a cool place to drain overnight, then squeeze as much juice as you can.
2. Place the juice into a large jar along with a spoonful or two of the pulp. Add the cider yeast (following the manufacturer's instructions) and give it all a good stir.

3. Place a piece of muslin or a fancy crochet cover over the top of the jar, seal with an elastic band and leave to ferment at room temperature for at least four weeks. Give the mixture a swirl now and then and remove any foam by skimming it off the top.
4. Once the liquid tastes sour and vinegar-like it is ready to go. Strain the liquid, or leave it with the 'mother' intact, into a sterilised bottle with a tight-fitting lid or cork. Add some of this vinegar to your next batch and it should speed the process along nicely.

CURES FROM THE KITCHEN

Help to soothe minor complaints, ills and chills with everyday ingredients you will find sitting on your kitchen shelves. The practice of using food to treat illness goes back to ancient times. Hippocrates, the father of medicine, spoke the wise words, 'let food be thy medicine and medicine be thy food' – something we often forget today. So consider reaching for the honey or salt rather than the tube of chemicals sold as a cure-all.

SALT

For athlete's foot, soak your feet in a warm saline solution daily until healed.

TONIC WATER

A small glass of tonic water before bed may help leg cramps and restless legs due to the quinine in the drink.

LEMONS

These citrus fruits are a natural disinfectant. Use lemon to lighten freckles and age spots, especially those on your hands. Rub a lemon half over the affected area and leave to dry.

HONEY

Honey is great for treating cuts and wounds as it has amazing antibacterial properties and will dry to form a natural bandage.

OLIVE OIL

Use it to treat dry itchy skin. Soak an ear bud and gently apply to your ear for minor earache.

OATS

Add to the bath for dry or irritated skin.

CAYENNE PEPPER

Mix in a little oil to form a paste. Apply the paste to chilblains or muscle aches and wrap in a dry cloth.

ONIONS

Mix 1 teaspoon of onion juice and 1 teaspoon of honey. Taken three times a day, this mixture can stave off a cold.

APPLE CIDER VINEGAR

Apply to bruises as an excellent anti-inflammatory. Sip 2 teaspoons in hot water to treat a cough.

BICARBONATE OF SODA

1 teaspoon in a glass of water can help to ease heartburn.

GINGER

Ginger soothes an upset stomach, and is great for relieving nausea, including travel and morning sickness. Sip an infusion of sliced ginger in hot water.

GARLICKLY GOODNESS

So much more than a cooking ingredient, garlic has many uses for health and also around the home. A great immunity booster, it is a powerful defence against bacteria, fungus, viruses and mould, and taken daily has been reported to protect folk from the common cold. Although it's pungent and powerful aroma is not appealing to everyone, its benefits far outweigh its smell, so don't dismiss it straight away and at least give it a try.

USE IN THE GARDEN AS A PESTICIDE SPRAY. Crush 3-4 cloves of garlic in a cup of hot water and let them sit for a day or two, strain, and then add to a spray bottle with 1 teaspoon of washing-up liquid and spray onto infested plants. Especially good for aphids and whiteflies.

TO HELP COMBAT ATHLETE'S FOOT, crush 2–3 cloves of garlic along with a tablespoon of vinegar into a bowl of warm water and soak your feet for 20 minutes.

OFF TO DO A BIT OF FISHING ...? Use garlic as bait. Coat your normal bait in garlic powder and the fish will swim straight for your hook.

RUB A CLOVE OF GARLIC ON A COLD SORE to help speed up the healing.

HELP REPEL MOSQUITOES AND OTHER INSECTS by rubbing garlic oil onto your skin... best not leave the garden or see any visitors on that day, though.

REPAIR SURFACE SCRATCHES ON GLASS by rubbing over the sticky juice and then wipe away any excess. Garlic has some adhesive properties and will help strengthen the damaged area.

ADD A CHOPPED CLOVE OF GARLIC to your usual cleaning spray to make an instant disinfectant. Especially good when illness has raged through the home and you need an extra something to banish the germs and bacteria.

HERBAL HONEY

Although honey is quite delicious and beneficial to your general wellbeing on its own, adding various herbs and spices can be a great way to reap the gentle medicinal properties of the plants. As well as eating honey it can be used externally. For example, lavender-infused honey is delicious on ice cream but can also soothe a minor burn, and rose-infused honey can make a great 5-minute facemask. Sage and thyme infusion is great for a sore throat, while chamomile infusion in a glass of hot water with lemon makes a calming bedtime brew.

Try to use honey from a local beekeeper as it will most likely be raw and packed full of lovely stuff foraged by the bees from your local vicinity. But failing that, you can often find a county or country blend. Herbs can be easily grown at home. Always try to source organic spices for the greatest medicinal properties.

INGREDIENTS
Various fresh or dried herbs, single or mixed
blends – rosemary, thyme, borage, lavender,
rose petals, chamomile, mint, sage, edible
florals, and so on
Glass jar
Raw honey – preferably from a local beekeeper

1. Place your herbs, spices or florals of choice into the jar, don't pack them in too tightly as the honey will need to be able to move around the jar and fill all the spaces.

2. Pop the jar of honey in a bowl of warm water and leave it there until the honey becomes fluid and easily pourable.

3. Pour the honey onto the herbs in the jar. Give the mixture a good stir and pack everything down gently, pressing down the plant material. When the jar is full, leave it in a warm place for at least 14 days, giving it a bit of a shake every 2–3 days. You may wish to strain the honey before using it. If so, leave for at least three weeks before doing so.

DANDELION SYRUP

Most of us see this glorious yellow flower as an annoying weed, popping up in our flower beds in early spring, but to beekeepers, it is met with a sigh of relief as pollen and nectar start to flow. But it also means swarms may be on the cards. This recipe is as old as the hills and is sometimes known as May honey. It is a great vegan alternative to the real stuff. Use it to help soothe a sore throat or a dry cough – mix a spoonful of syrup with a squeeze of lemon juice in a mug and top up with hot water.

INGREDIENTS
650ml spring water
Grated rind and juice of 1 lemon
250g fresh dandelion flower petals
1.5kg granulated sugar
Sterilised glass jars

---

1. Place the water, lemon juice and rind and the dandelion flowers into a pan. Bring to the boil, then reduce the heat to low. Gently simmer for 30 minutes, then remove from the heat and allow to cool overnight. This allows the flavour of the petals to really infuse the water.

2. Strain the liquid through a fine sieve or muslin-lined colander and return to the pan.

3. Stir in the sugar and simmer for 15–20 minutes, until the sugar has dissolved and the mixture has thickened slightly. Pour into sterilised glass jars, allow to cool, and store in the fridge. Consume within the month…if it lasts that long.

GINGER COMPRESS

A stuffed-up nose is often the worst part of a cold, flu or hay fever. Homemade natural remedies can often be very effective at relieving a blocked nose. Because ginger helps to soothe inflammation and is such an all-round healing spice, it is a great choice for this compress.

INGREDIENTS
3 tbsp freshly grated ginger
Cupful of water
Large strip of muslin

1. Place the grated ginger and water into a saucepan and warm the mixture on the stove. Drain the liquid into a bowl and soak the strip of muslin.

2. When the liquid has cooled slightly, squeeze excess fluid from the cloth, fold it up and place it over your nose and forehead. Leave on until cool, then repeat until you feel some relief.

REJUVENATING BORAGE & HONEY TEA

Borage is a beautiful plant that is easily grown in any garden. Its blue star-shaped flower is a favourite of bees, and borage has a long list of medicinal properties. Taken as a tea, borage can help to rejuvenate you after an illness, and it can lift the spirits and bring the body and mind back into balance. This mixture is especially good for poorly children who need to build up their strength after an illness.

INGREDIENTS
10g dried borage flowers and leaves
500ml water
1 tsp honey per cup

---

1. Place the borage into a bowl.

2. Boil the water and let it cool for 5 minutes, then pour it over the borage and leave to infuse for 30–40 minutes.

3. Strain the liquid and add honey to sweeten. Sip a cup every couple of hours. You may wish to gently warm the tea before you add the honey.

BE STILL AND LISTEN TO THE BEES
FOR THEY ARE WISER THAN WE KNOW

HONEY & LEMON SOOTHER

As we slide into autumn and the season of coughs, colds and sniffles approaches, it is really handy to have a bottle of this simple soother stashed away in the fridge. Just three simple ingredients blended together can soothe a raspingly sore throat, ease a tickling cough or generally just give you a bit of comfort. Honey contains all sorts of wondrous things, which makes it a great fighter of bacteria and viruses. Lemon is jam-packed with vitamin C, which your body is crying out for when the bugs have marched in. Glycerin will soothe a sore throat. Store this mixture in the fridge for about six months, which will get you over that chilly season when we are all susceptible to falling ill.

MATERIALS
150ml glass bottle – an old medicine bottle is ideal
50ml raw honey – from a local source if possible
50ml food-grade glycerin
50ml fresh lemon juice – strained to remove
any bits or pips

---

1. Begin by making sure your glass bottle is clean and sterilise it either in some boiling water or by popping it in the oven at 150°C/gas mark 2 for 15 minutes.
2. Pour the honey, glycerin and then the lemon juice into the bottle while it is still warm and shake well to combine all the ingredients.

3. Keep the bottle in the fridge and use when needed. Shake well before pouring. Take 1–2 teaspoons every couple of hours until you begin to feel better. Or add it to a cup of warm water for a soothing drink.

KEEP WELL SYRUP

A blend of foraged hedgerow fruits makes an excellent syrup to take daily to ward off winter bugs and colds. Rosehips, elderberries, blackberries, rowan and hawthorn all add a little something to this little bottle of goodness. For a real winter warmer, add in some optional spices, a slice or two of orange, a bit of ginger, clove or cinnamon and hot water and you have a tasty drink to warm you right down to your toes.

INGREDIENTS
500g mixed hedgerow fruits
Grated rind of ½ lemon
500ml boiling water
Jelly bag
Granulated sugar
Spices (optional) – 1 cinnamon stick,
2 all-spice berries and 3–4 cloves

1. Wash your fruits thoroughly, making sure you have removed any debris or small insects that may have tagged along. Give them a quick blitz in a food processor.

2. Place the fruits in a large pan along with the grated lemon rind and optional spices, and add the boiling water. Cover the pan with a lid and bring back to the boil, then reduce the heat and simmer for 10 minutes.

3. Remove from the heat and allow the mixture to stand until cool.

4. Pour into a jelly bag and leave for several hours until the liquid has drained through.

5. Measure the liquid, and then return it to the saucepan. For every 500ml of juice, mix in 500g granulated sugar. Heat gently until the sugar has dissolved, then pour it into sterilised bottles and seal. Consume within three months. Keep opened bottles in the fridge and use within a week of opening.

CARAWAY TEA FOR A JIPPY TUMMY

Caraway tea has been used for generations to cure all manner of ills, from bronchitis to colic, but it is best known for its gentle soothing action on the digestive tract. Sip a warm cup of caraway tea to help ease an upset stomach or take a cup after your meal to help settle any uncomfortable wind or bloating. Caraway seeds are powerhouses of micronutrients, so adding a cup or two of this brew into your daily diet can be beneficial. Below are two methods for making the tea – a quick one-cup method, and a longer cold-fusion method that produces a tea you can keep in the fridge for a week at a time. The latter is ideal if you wish to drink this tea daily.

INGREDIENTS

Caraway seeds

Boiling water

Jar (for overnight brewing)

ONE-CUP METHOD

1. Place 1 tablespoon of caraway seeds into a cup and cover with boiling water. Cover to keep the heat in and allow the seeds to steep for 15–20 minutes.

2. Strain into another cup and sip slowly. Add honey to taste if you like a sweeter brew.

OVERNIGHT BREW

1. Place 4 tablespoons caraway seeds in a heatproof jar and fill with boiling water. Fix the lid on the jar and allow to cool. When cool, place in the fridge and leave overnight.

2. Strain the seeds from the brew and re-jar. Keep in the fridge for up to a week. To use, dilute to taste with hot or cool water, depending on your preference. Add honey for sweetness.

LEMON BALM OINTMENT

Feeling run down, with the familiar tingle that means there's a cold sore outbreak on the way? Nip it in the bud with this effective ointment and help speed up the healing process to banish those pesky cold sores. This ointment can also be used to give some relief to chicken pox or shingles sores, or as a general healing salve to soothe irritated skin.

INGREDIENTS
30ml almond oil

1 tbsp beeswax

1 tbsp cocoa butter

10 drops melissa or melissa-blend
essential oil (see below)

5 drops tea tree essential oil

5 drops lavender essential oil

2 x 15ml pots

---

1. Place the almond oil, beeswax and cocoa butter into a bowl set over a pan of simmering water and stir until the solids have melted and are well combined. Allow the mixture to cool slightly.

2. Add the drops of essential oils into the base mix, stirring constantly, then pour your ointment into the small pots and secure the lids tightly.

3. Pop the pots into the fridge to allow the ointment to harden. When a cold sore starts to tingle, apply the ointment every couple of hours until the cold sore has healed. If kept properly sealed, the ointment will keep for at least a year.

MELISSA-BLEND OIL
Pure melissa oil is rare and costly and often adulterated with various other cheaper oils such as lemongrass, lemon verbena or lemon. A good blend will be very similar in composition to the pure oil, but at a much more affordable price. So if you feel like splashing out on a bottle of true oil, buy it from a reputable company that has information about source and extraction processes available on their oil products, so you can be assured that you are getting what you paid for.

FRAGRANT BATH TEAS

There is nothing quite like a soak in a hot bath to ease away the dirt & worries of a tough day. Why not add a little natural goodness to your bath to help sooth and naturally heal your aches and pains.

Following the recipes below or make up your own from your favourite herbs and florals, mix the dry ingredients together and them fill a reusable muslin tea bag.

Place the bag into a bath of hot water and let it infuse for a minute or two before you climb in and soak away the day.

GOOD FOR ACHES & PAINS

2 tablespoons dried lavender flowers

2 tablespoons dried rosemary

2 tablespoons chamomile flowers

2 tablespoons Epsom salts

WAKE UP CALL

2 tablespoons dried mint leaves

2 tablespoons dried basil

2 tablespoons dried orange peel

2 tablespoons dried rosemary

REST AND RELAX

2 tablespoons dried rose petals

2 tablespoons dried chamomile flowers

2 tablespoons dried lavender flowers

2 tablespoons oatmeal

SOOTH A WEARY HEAD

2 tablespoons dried calendula petals

2 tablespoons dried lemon balm

2 tablespoons dried lavender flowers

2 tablespoons oatmeal

FIND PEACE IN THE
ORDINARY THINGS

BUTTERMILK WRAP FOR SUNBURN

It is always best to avoid too much exposure to the sun and there are many lotions and potions that can protect you from those harmful rays. Unfortunately, they are often full of chemicals, too. So if you get caught out and suffer the dreaded burn, to avoid slapping on more chemicals in the form of aftersun, why not try a buttermilk wrap to ease the pain and help speed up the healing of the burnt skin? Lavender oil is well documented for its treatment in skin burns and will give some pain relief along with the cooling effect of the buttermilk.

INGREDIENTS
200ml buttermilk or natural live yogurt
Few drops of Lavender essential oil
Piece of cotton muslin

1. Place the buttermilk, or yogurt, or a blend of the two, into a large bowl and add 3–5 drops of lavender essential oil. Stir really well until everything is well mixed together.

2. Fold your piece of cotton muslin into a manageable size, then soak it in the buttermilk mixture. Apply to the burned area and leave for 10–15 minutess or until the wrap has become warm. Repeat if necessary, then wash the skin with cool water. You can continue this treatment every couple of hours until the redness calms and the skin has begun to heal.

natural BEAUTY

NATURAL BEAUTY IS THE CROWN IN THE HOMEMADE LIFESTYLE, YET IT IS SOMETHING THAT IS OFTEN OVERLOOKED. ADDITIVE-FREE POTIONS AND LOTIONS CAN ONLY BE OF BENEFIT TO YOUR HEALTH AND WELLBEING, AND THE MORE YOU REDUCE THE OVERLOAD OF STUFF THAT IS INGESTED OR SOAKED INTO YOUR SKIN, THE EASIER IT WILL BE TO ACHIEVE THAT HEALTHY GLOW. CREATE YOUR OWN CLEANSING AND TONING LOTIONS, HELP BANISH PESKY SPOTS OR TREAT YOUR HAIR WITH AN EGG OR TWO WHILST RELAXING WITH A REJUVENATING SUNFLOWER SEED MASK. DRIFT OFF INTO A CALM STATE AND LET YOUR HEART SING.

ALMOND & ROSEWATER CLEANSING CREAM

A gentle blend of almonds and rosewater made into a cleansing cream is a great homemade addition to your daily skincare routine. As a cream it effectively removes dirt and grime but it also softens your skin. Follow it with the lavender and cider vinegar toner (see page 129) and your skin will be thankful for the natural approach to its care and reward you with a beautiful radiant glow.

INGREDIENTS
60ml almond oil
1 tbsp beeswax
2 x 100iu Vitamin E capsules
60ml rosewater
1 heaped tbsp ground almonds
Glass jar

---

1. Warm the almond oil and beeswax in a double boiler until the beeswax has melted.
2. Remove from the heat and mix in the Vitamin E oil from the capsules. Gently whisk the oils as you slowly add in the rosewater and keep whisking until all the rosewater has been incorporated and the mixture turns to a creamy white lotion.

3. Stir in the ground almonds, then pour the mixture into a jar. Put it in the fridge to cool. Use daily to cleanse away the dirt and grime left behind by the day. Rub a small amount across your face with your fingers, then remove it with a cotton pad. Follow with a suitable toner to remove oily residue. Keep in the fridge and use within 1 week.

LAVENDER

Lavender is one of the most prized herbs. Its sweet-smelling foliage and deep violet hues can lift the most down of hearts. Uses range from hair rinses, skin care, keeping your linens smelling sweet and pest-free to gently flavouring a delicate biscuit or your favourite tipple. Essential first aid for burns, bites and stings, lavender indeed takes the number one slot and is a queen amongst plants.

A FEW DROPS of lavender oil applied directly to a burn can speed healing time and lessen any pain.

CRUSH FRESH FLOWERS and rub onto your skin to help ward off the pesky midges and mosquitoes.

LAVENDER OIL RUBBED onto your temples can ease a tension headache.

A DROP on a small person's pillow can help them drift off into a dreamy sleep.

ADD 1 TEASPOON lavender flowers to homemade lemonade and serve over ice for a refreshing summer drink.

USED AS A HAIR RINSE, lavender can soothe an inflamed scalp, balance oil production and help to prevent hair loss.

MIX A TABLESPOON of lavender flowers with a cup of bicarbonate of soda and sprinkle onto a carpet or rug, leave for 30 minutes (or ideally overnight) and then vacuum for a fresh-smelling carpet.

PLACE BUNCHES of dried lavender flowers in cupboards to deter moths and silverfish.

PLACE A TEASPOON of lavender flowers into a jam jar of caster sugar and leave in a dark cupboard for a few weeks, then use to sprinkle on cakes and biscuits, or even your porridge, for a gentle floral treat.

INFUSE HONEY with lavender flowers, see page 98.

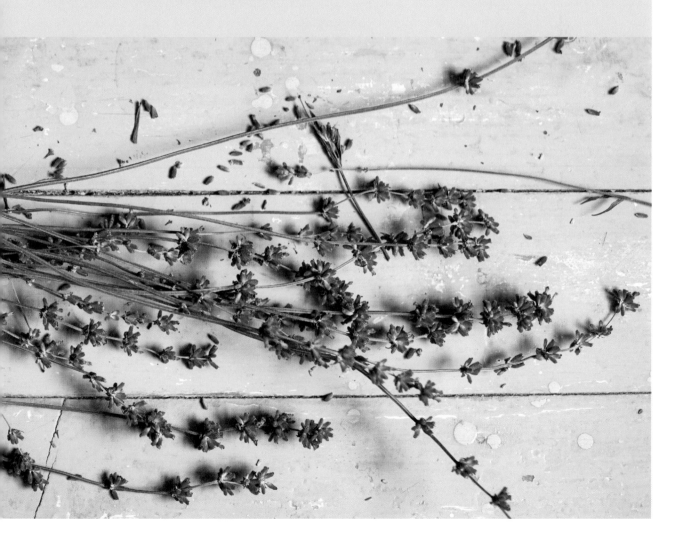

BANISH ACNE SKIN TONIC

Spots and blemishes are unwelcome guests that often arrive when least wanted. Unfortunately, stressful days, hormone imbalances and the wear and tear of the everyday can mean that many people suffer with them at various times of their lives… they aren't just the bane of the teenage years. If an outbreak occurs, make up a small batch of this tonic to dab onto the offenders. It will help speed the healing process and get your skin back to its rosy old self.

INGREDIENTS
40ml freshly brewed green tea
1 tsp grated ginger
20ml apple cider vinegar
15ml glycerin

1. Brew a cup of green tea and measure 40ml into a small bowl. Add the grated ginger. Leave to infuse until cool.

2. Strain the liquid into a small bottle and add the apple cider vinegar and glycerin, shaking well to combine all the liquids together. Using a cotton bud or similar, dab onto the spot every couple of hours until healed. This mix will last 1–2 weeks in a cool dark place. Make a new batch up as needed.

HONEY SOAP

There is something quite rewarding about making your own soap, especially as you rub a bar between your hands and the soft white bubbles float across your skin. This recipe makes a gentle all-purpose bar with a little bit of honey added to give it an extra-special something.

INGREDIENTS

2 tbsp honey

400ml water

160g sodium hydroxide (caustic soda)

200g coconut oil

950g olive oli

Soap moulds

---

1. Dissolve the honey in a little hot water.

2. Make the lye solution. Fill a glass jug with the measured water and add your sodium hydroxide to it, stirring carefully with a wooden spoon until it is dissolved. Set to one side.

3. Combine the oils in a large stainless-steel pan and heat gently. Once they are melted, remove the pan from the heat.

4. Once the temperature of the oils and lye are between 50°C and 37°C, place your stick blender into the oil and start to blend. While blending, pour the lye solution into the oil mixture, being careful not to splash while mixing.

5. Keep blending until the mixture traces. (Tracing looks like a thickened sauce and will show your stir marks for several seconds.)

6. Stir in the honey and water.

7. Pour the soap mix into your moulds and leave to set for 24 hours.

8. Remove the soap from the moulds and cut into bars, if needed, then lay them out to cure. Leave them for six weeks to complete the saponification process.

A NOTE ON MOULDS

You can use all sorts of things for soap moulds, including silicone bakeware, clean milk cartons, plastic tubs … the list goes on. Just remember that you have to remove the soap from them, so make sure the shape is practical for enabling you to do so.

SUNNY SUNFLOWER SEED MASK

This facemask contains all manner of skin-saving antioxidants. Sunflower seeds, packed full with vitamins A, D and E, as well as many other beneficial natural compounds, will help to gently smooth your skin, banish those lines and wrinkles and make the outer you shine a little bit brighter. The honey brings its anti-inflammatory, antiseptic and antibacterial properties to the mix, which means that it will help to ward off any spots or pimples that may be unwelcome guests. Cheap and easy to make, a weekly treatment is an absolute essential for anyone's skin.

INGREDIENTS
½ cup sunflower seeds, ground to a powder
(use a food processer or coffee grinder for this)
1 tbsp honey
1 tsp almond oil
1 tbsp warm water

1. Mix all the ingredients together into a paste.
2. Apply to a freshly washed face and relax for 20 minutes.

3. Wipe away with a damp cloth and then rinse face.

TAKE A DEEP BREATH AND
BEGIN ANYWHERE

SOLID BEESWAX PERFUME

If you have ever fancied dabbling in perfume making, this is the ideal project to get you started. The roots of homespun alchemy lie in the stuff of many childhoods – sneaking bits and pieces from the garden and mixing them in a jam jars, adding the inside of an old felt tip for colour and hoping a magical scent would be created instead of the brown sludge that always followed. Use an old pill box, compact or tiny tin and give your carefully blended scents as gifts.

INGREDIENTS
small container with lid
1 tsp unscented beeswax
4 tsp jojoba or sweet almond oil
small jam jar
wooden skewer
40 drops essential oil blend

---

1. Prepare a small container that has a lid, ensuring it is clean and to hand.

2. Place the beeswax and oil into a small jam jar and place in a saucepan.

3. Fill the pan with water so that the level of the water rises just above the level of the contents of the jar. Heat gently, stirring the mixture with a wooden skewer.

4. When all the beeswax has melted into the oil, remove the jar from the saucepan and continue to stir until the mixture begins to cool and thicken.

5. Quickly add your oil blend, stirring continuously.

6. Pour the mixture into your prepared container, put on the lid and leave to cool completely. Leave for 1–2 weeks in a cool, dark place to allow the scents to develop. Use within six months, as the scent will begin to fade once it has been opened.

OIL BLENDS TO TRY

Mix oils together in a small amber bottle and keep in a dark place until ready to use.

ADD A SPRING TO YOUR STEP

20 drops frankincense and 20 drops grapefruit.

A HEART WARMING HUG

20 drops geranium, 15 drops of lavender and 5 drops black pepper.

FRESH AS A DAISY

10 drops orange, 10 drops of lemon, 10 drops of palmarosa and 10 drops cedarwood.

HOMEMADE INFUSED ESSENTIAL OILS

Although not as potent as distilled essential oils, these are a great cheaper alternative to have in the cupboard and can be put to all sorts of uses. These infused oils can be applied to cuts and bruises, and are great to include into your cleaning routine to keep various creepy crawlies out of the house. They have a shelf life of about a month or so if kept in a dark, cool place. Experiment with using oils infused with different herbs and spices around the house. Do remember some plants will give up their scents far more easily than others. There is a reason why rose, jasmine and melissa oils are so highly priced.

INGREDIENTS
Herbs or spices (see below)
200ml grapeseed oil
2 glass jars with tight fitting lids

---

1. Place your herbs or spices in a large glass jar with a tight fitting lid and cover with the grapeseed oil.
2. Place the jar into a pan of water and gently heat to 65°C. It is important that the temperature of the water reaches 65°C exactly as, at this point, the molecules containing the smelly goodness burst and release themselves into the oil. Keep at this temperature for 5 minutes, then remove the pan from the heat and leave it to cool.
3. Remove the lid and strain the oil into a clean glass jar. Store in a dark glass jar and use as needed.

HERBS AND SPICES TO USE
Experiment with combinations of the plants that take your fancy, or are growing rampantly in your garden. The following suggestions should get you started:

FOR LEMON OIL, use the grated zest of 5 unwaxed lemons.

FOR CLOVE OIL, use 2 tbsp whole cloves (great for banishing spiders)

FOR ROSEMARY AND THYME OR LAVENDER OIL, use enough leaves and stalks to fill-half your jar.

LAVENDER & CIDER VINEGAR TONER

An excellent toner for the skin, this blend of lavender and apple cider vinegar makes a great addition to your skincare regime. Both ingredients are renowned for their antiseptic and antibacterial properties, which makes them ideal to use on problem areas, and apple cider vinegar (see page 94) will help to restore the natural pH balance of your skin, making it super soft with a rosy glow.

INGREDIENTS
50ml apple cider vinegar
100ml lavender hydrosol
Small bottle

1. Pour the ingredients into a small bottle and shake well. Use morning and evening as part of your skincare routine. If you find the vinegar smell to be a little over powering then dilute with more hydrosol.

MAKING YOUR OWN PLANT HYDROSOL
You can make your own plant hydrosols using an old stove-top Italian-style moka coffee pot. Place your plant material in the base of the pot and cover with water. Remove the filter where the ground coffee would go and screw on the lid. Heat gently on the stove until the hydrosol has gathered in the top of the coffee pot, then pour it into a heatproof jar and place the lid on immediately. When cool, store it in the fridge and use as needed. Discard any unused hydrosol after a month and make a fresh batch. It is advisable not to use the pot to make coffee in again, so treat yourself to a new pot and keep them separate.

CARE WHEN USING ESSENTIAL OILS

A cautionary tale… Although it is unusual for people to have a reaction to natural ingredients, it does happen, so be wary when using something new and always test it out on a little hidden patch of skin before you cover yourself from head to toe in your newly made concoctions. Follow these few simple steps and hopefully you should be a-okay. If you do have a reaction, stop using the offender straight away and steer clear of those ingredients in the future.

ALWAYS DILUTE ESSENTIAL OILS – never use them neat.

ALWAYS TEST natural beauty products on a patch of hidden skin. Leave for 24–48 hours to check for a reaction.

STORE ESSENTIAL OILS CAREFULLY, and somewhere dark and cool. Always keep the tops screwed on tightly to prevent evaporation.

IF ANYTHING SMELLS ODD, discard it and make up a new batch.

ALWAYS STERILISE any jars or bottles before using them.

BUY FROM A REPUTABLE SELLER to ensure that you are getting the purest oil and cosmetic grade ingredients.

HOLD ON, LET GO, EVERYDAY
IS ABOUT TRYING TO BALANCE

NETTLE & CHAMOMILE HAIR RINSE

To have glowing locks of healthy hair is indeed a wonderful thing, something that all of us without it wish for each and every day. Itchy scalps and troublesome dandruff can drive you mad and often the only treatment on offer can be harsh on your hair and leaves everything looking dull as dishwater. This recipe contains nettle to help with pesky dandruff and the chamomile will help to soothe your scalp and brighten your hair.

INGREDIENTS
1 tbsp dried stinging nettles
1 tbsp dried chamomile flowers
100ml boiling water
1 tbsp apple cider vinegar

1. Place the nettles and chamomile in a bowl and pour over the boiling water. Leave the herbs to infuse until the liquid is cool, then strain.

2. Pour the strained liquid into a jar or bottle, add the cider vinegar and give it a good shake. Massage through your hair after shampooing and leave on for 5 minutes before your final rinse with water.

HERBAL HAIR RINSES

Use natural herbal infusions to nourish and promote a healthy scalp, soften your hair and help treat conditions such as dandruff or a sensitive skin, while helping to restore shiny bouncy locks. To prepare, place two cups of your chosen herb into a bowl and cover with boiling water. Allow the 'tea' to steep for at least 30 minutes, strain it and allow to cool. Use as your final hair rinse, dry hair as usual.

BLACK TEA – helps to prevent shedding and can darken those early grey hairs.

CALENDULA FLOWERS – will soothe a sensitive scalp and help protect it from bacteria while adding a warm glow to your locks.

DANDELION LEAVES – help to treat dandruff and dry hair, and will also add a sheen to dull locks.

GINGER – add a bit of zing to your scalp. Stimulates hair follicles and encourages growth.

LEMON BALM – great for oily hair. A mild astringent, it will help to wake up and refresh the scalp whilst balancing oil production.

NETTLE – help to prevent hair loss, stimulate the scalp and encourage glossy shiny hair. Can also curb dandruff and split ends, so pretty much a great all-rounder.

MINT – its antibacterial and astringent properties can help to increase blood flow to the scalp and stimulate hair growth.

ROSE-HIPS – jam-packed with vitamin C, they help to strength the hair and also give a boost to red hair.

ROSEMARY – a revitalising herb that will improve and encourage hair to be strong and shining. Can help with thinning hair by stimulating new growth.

SEAWEED – can add a bit of bounce to limp and lifeless hair, as it's full of minerals and proteins that offer the roots strength and vitality.

YARROW – great for an itchy scalp, it helps reduce excess oil and soothe any sensitivity. Regular use can lighten hair.

EGG HAIR TREATMENT

Many people suffer from dry hair from time to time. Sun, sea, water, a day in the garden or the rigours of chemical hair dyes and styling products can all take their toll, damaging your hair and turning it into the dry, straw-like mop that leads to much despair. Regular conditioning treatments can help at such times but, rather than reaching for a pot of something expensive, why not try this age-old remedy with just a few items from the kitchen and see if it does return the lustre to your locks.

INGREDIENTS
1 egg yolk
1 tbsp melted coconut oil
1 tbsp olive oil

1. Carefully separate the egg yolk from the white of the egg and place it in a small bowl.

2. Gently warm the coconut oil with the olive oil and then drizzle it into the yolk as you whisk the mixture.

3. Coat your hair with the mixture and massage it in. The mixture will become a little stiff as the coconut oil cools down.

4. Cover with a towel and leave on for about 30 minutes.

5. Shampoo and rinse well. Style and dry as normal, admiring your shiny locks as you do.

LEATHER NAIL BUFFER

Dare to go bare – ditch the nail varnish full of chemicals and leave your nails natural. Regular buffing will keep your nails looking healthy and maintain their natural shine. Encouraging blood to flow to your fingertips could help your nails grow and you will have beautiful shiny nails. Commercial nail buffers should only be used sparingly as they can weaken the nail surface, so why not make one that you can use every day and keep your nails beautiful and healthy?

INGREDIENTS
Small piece of chamois leather
Stuffing – wool is ideal but polyester
toy filling is also suitable
Strong thread

(Use a 1cm seam allowance throughout.)

1. Cut a rectangle measuring 12 x 10cm from a piece of chamois leather. Fold it in half along the 12cm length and stitch down the long side.

2. Turn seam to the inside and bind thread around one end.

3. Fill with stuffing, packing it in well so that the buffer is firm

4. Leave a couple of centimetres at the top edge, bind thread around the end and knot to seal the stuffing in. Trim off any excess at either end. Buff away until your nails shine.

BEESWAX NAIL CREAM

Our hands are in constant use and can easily become dry and chapped, especially throughout the colder months of the year. Likewise, particularly during winter, nails and cuticles often split and tear making for rather sore fingers. Using a little of this sweet-smelling beeswax nail cream on your nails and cuticles every time you buff them will both condition and strengthen them.

INGREDIENTS

15ml almond oil

½ tbsp beeswax

15ml glass jar

6 drops lemon essential oil

1. Place the almond oil and beeswax into the glass jar and pop it into the microwave. Heat gently in 5 second bursts until the beeswax has melted. Allow the mixture to cool slightly. Alternatively, you can pop the lid on the jar and place it in a shallow bowl of boiling water until melted. You may have to change the water if it cools before melting the wax.

2. Add the drops of essential oil into the base mix, stir to combine, then put on the lid and secure it tightly. Pop the jar into the fridge to allow the nail cream to harden. Rub a little into each nail before you buff.

suppliers and interesting information

Many of the items needed to complete the projects in this book can be found in local hardware shops, health food stores and on the supermarket shelves. If you are having trouble sourcing something do ask your local shopkeeper as they can often get you a good deal on bulk items such as vinegar, bicarbonate of soda, castile soap, etc. There are some online sources below but try to use these as a last resort and keep local shops alive.

HOUSE & HOME

For beeswax, arrowroot, bicarbonate of soda and other bulk items
www.dockandnettle.com
www.realfoods.co.uk

Organic cotton ideal for beeswax wraps, tote & produce bags
www.organiccotton.biz

Jars & bottles
www.kilnerjar.co.uk

Weck jars
www.utilitygreatbritain.co.uk
www.foalyard.co.uk

Willow
www.willowwithies.co.uk

Lambswool stuffing
www.worldofwool.co.uk

Saving energy
www.saveenergy.co.uk
www.defra.gov.uk
www.greenshop.co.uk

wild food
www.wildfood.info

NATURAL HEALTH & BEAUTY

essential oils
www.essentialoilsonline.co.uk
www.avena.co.uk
www.quinessence.com

Natural ingredients & herbs
www.baldwins.co.uk

Cider yeast
www.welovebrewing.co.uk

Herb seeds and info
www.jekkasherbfarm.com
www.herbpatch.co.uk

Candle wax
www.candlemakers.co.uk
www.candle-shack.co.uk

index

A

aches, fragrant bath teas 112
acid reflux 84
acne 85
    banish acne skin tonic 120
age spots 96
all-purpose orange cleaner 20
almond and rosewater cleansing cream 116
antibacterial: anti-bacterial spray 20
    remedies 96
antiseptics 85, 94
apple cider vinegar 65, 89, 96, 120, 129
    homemade 94
appliance cleaner, kitchen 21
apron, harvest 46–8
athlete's foot 96, 97
autumn produce 80, 81

B

bags: cotton produce bags 49–51
    market day tote bag 37–9
basket, wire laundry 33–4
bath teas, fragrant 112
beauty, natural 114–37
bedding, bits and bobs eiderdown 54–6
bees 82
beeswax: beeswax furniture polish 14
    beeswax nail cream 137
    reusable beeswax food wrap 40–1
    solid beeswax perfume 125
berries 73
bicarbonate of soda 21, 27, 35, 43, 53, 96
birds 73
    bird feeder bottle 86
bite relief 65, 118
bits and bobs eiderdown 54–6
blankets 58
    fabric-backed blankets 60
blood stains 28
borage, rejuvenating borage and honey tea 105
borax 29
bottles, care of 42–3
brass, shining 93
bruises 96, 126
buckets, spring clean 17
bugs, preventing 24, 70
burns 85, 118
buttermilk wrap for sunburn 113

C

calcium supplements 65, 72
calendula herbal hair rinse 132
candles: beeswax eggshells 65
    candle in a jar 57
caraway tea for a jippy tummy 109
carpet cleaning 119
cayenne pepper 96
chamomile, nettle and chamomile hair rinse 131
chemicals: and bees 82
cleaning without 20–1
cherries 73
chickens 65, 89
chilblains 96
cleaning 12–25
cleansing cream, almond and rosewater 116
clothes 58
clover, white 82
coffee 72
    coffee stains 28
cola stains 28
cold remedies 93, 96, 102, 108
    honey and lemon soother 106
cold sores 97, 110
compost 65, 72, 77
compress, ginger 102
copper: cleaning copper pans 25
    copper tape 73
    shining 93
cornflour 21, 28
cotton: cotton produce bags 49–51
    softening cotton blankets 93
    washing 29
coughs 101
    honey and lemon soother 106
cramp 96
cures from the kitchen 96
curtains 58
cuts 96

D

dandelions 82, 132
    dandelion syrup 101
decal plant markers 78
decanters, cleaning 22
diffusers, homemade reed 66
disinfectant 96, 97
dogs 93
draught excluders 58
drying washing 58
    wire laundry basket 33–4
    wool dryer balls 30

E

earwigs 73
eggs and eggshells 65, 89
    egg hair treatment 134
    soil improvers 72
    use in laundry 29
eiderdown, bits and bobs 54–6
energy saving tips 58
Epsom salts 72
essential oils: care when using 130
    homemade infused 126
    homemade reed diffusers 66

F

fabric: fabric-backed blankets 60
    fabric fresh spray 35
facemasks, sunny sunflower seed 122
fertilisers, natural 72, 82–3
fingernails, whitening 25
fires, kindling wraps 63
fishing bait 97
floor cleaner, forest-fresh 18
flu remedies 102
food: reusable beeswax food wrap 40–1
    seasonal 80–1

forest-fresh floor cleaner 18
fragrance: fragrant bath teas 112
   homemade reed diffusers 66
   homemade smudge sticks 59
freckles 96
fridges, cleaning 93
frogs 73
fruit 80
   removing fruit juice stains 28
fruit flies 93
furniture: beeswax furniture polish 14
   furniture wipes 20
   wood wipes 13

G
gardens 68–89
garlic 97
ginger 96, 120, 132
   ginger compress 102
glass: cleaning 22, 65
   repairing scratches 97
grapefruit, slug control 73
grapefruit seed oil 89
grapeseed oil, homemade infused essential oils 126
grass stains 28
grease stains 22, 28
green tea, banish acne skin tonic 120

H
hair care 85, 119
   egg hair treatment 134
   herbal hair rinses 132–3
   nettle and chamomile

hair rinse 131
hangover cures 85
harvest apron 46–8
hay fever 85, 102
headaches 118
health, natural 90–113
heartburn 96
heating, energy saving tips 58
hedgehogs 73
hedgerow fruits, keep well syrup 108
herbs 70
   compost booster 77
   herbal hair rinses 132–3
   herbal honey 98
   homemade smudge sticks 59
   willow wire herb hanger 45
honey 85, 96
   herbal honey 98, 119
   honey and lemon soother 106
   honey soap 121
   rejuvenating borage and honey tea 105
   sunny sunflower seed mask 122
hot water bottles 58
household stain powder 27

I
indigestion 85
insects 73, 97
invisible ink 24
itchy skin 96

J
jars: candles in 57
   care of 42–3

K
keep well syrup 108
kettles 58, 93
kilner jars 43
kindling wraps 63
kitchens 36–51
   kitchen appliance cleaner 21

L
labels, removing 43
laundry 26–35, 58, 65
lavender 70, 118–19
   lavender and cider vinegar toner 129
leather: cleaning leather shoes 24
   leather nail buffer 136
leg cramps 96
lemon balm 70, 133
   lemon balm ointment 110
lemons 24–5
   basic stain-removal kit 28
   cleaning glass 65
   disinfectant 96
   honey and lemon soother 106
   lemonade 25, 119
   use in laundry 28, 29
light bulbs, energy saving 58
limescale 22, 25, 93
linen, washing 29
lip balm 85

M
market day tote bag 37–9
mason-style jars 43
methylated spirit 28
microwaves, cleaning 24

midges 118
milk 28
mirrors, cleaning 22
morning sickness 96
mosquitos 97, 118
moths 119
mud stains 28
muscle aches 96
musty smells 43

N
nail care: beeswax nail cream 137
   leather nail buffer 136
nail polish stains 28
nausea 96
nettles: herbal hair rinse 133
   nettle and chamomile hair rinse 131

O
oats 96
oil stains 28
ointment, lemon balm 110
olive oil 43, 96
   egg hair treatment 134
   honey soap 121
onions 96
oranges, slug control 73

P
pains, fragrant bath teas 112
paint, homemade 53
pans: cleaning copper 25
   energy saving tips 58
parsley 70
pen stains 28
peppermint 70
perfume, solid beeswax

125
pesticides 97
pests 70, 73, 97
pets 93
pewter, shining 93
pick me ups 25
picture frames, cleaning 22
plant food 65, 76, 93
plant markers, decal 78
polish, beeswax furniture
    14
potatoes, removing stains
    28
produce bags, cotton
    49–51
pumpkin seeds 89

R
reed diffusers, homemade
    66
rejuvenating borage and
    honey tea 105
rest and relax fragrant bath
    teas 112
reusable beeswax food
    wrap 40–1
rice, preventing it sticking
    25
rose hips, herbal hair rinse
    133
rosemary 70, 133
rosewater, almond and
    rosewater cleansing
    cream 116

S
salad, windowsill 74–5
salt 96
    basic stain-removal kit
    28
    household stain powder
    27
saucepans: cleaning copper
    25

energy saving tips 58
seasonal produce 80
seaweed 72, 133
shoes, cleaning leather 24
silverfish 119
skin care 96
    almond and rosewater
        cleansing cream 116
    banish acne skin tonic
        120
    honey soap 121
    lavender and cider
        vinegar toner 129
    sunny sunflower seed
        mask 122
sleep 118
slugs 65, 72
smudge sticks, homemade
    59
soap, honey 121
soil, nourishing naturally
    72
sooth a weary head
    fragrant bath teas
    112
sore throats 101
sprays: anti-bacterial 20
    fabric fresh 35
spring clean bucket 17
spring produce 80, 81
stains 28, 29, 93
    basic stain-removal kit
    28
    household stain powder
    27
sticky residue, removing
    43
stomach upsets 85, 96
    caraway tea for a jippy
        tummy 109
sugar, lavender 119
summer produce 80, 81
sunburn, buttermilk wrap
    for 113

sunflower seed mask,
    sunny 122
supplements, calcium 65
sweat marks 28

T
tea: caraway tea for a jippy
    tummy 109
    herbal hair rinse 132
    rejuvenating borage and
        honey tea 105
thermostats 58
toads 73
toilet cleaner 21
tomatoes 65, 72
toner, lavender and cider
    vinegar 129
tonic water 96
tote bag, market day 37–9
travel sickness 96

V
vases, cleaning 22
vegetables 65, 72, 80
    windowsill salad 74–5
vinegar: apple cider
    vinegar 89, 94, 96,
    120, 129
    cleaning with 13, 18, 20,
    21
    and natural health 93
    use in laundry 28, 29, 35

W
wake up call fragrant bath
    teas 112
washing soda 18, 20, 29
weeds 82, 93
whites, keeping white
    29, 65
wild food 81
willow wire herb hanger
    45
windowsill salad 74–5

winter: energy saving
    tips 58
    winter produce 80, 81
wipes: furniture 20
    wood 13
wire laundry basket 33–4
wood: kindling wraps 63
    wood wipes 13
    woodash 72
wool blankets, softening
    93
wool dryer balls 30
wounds 96

First published in Great Britain in 2015 by
Kyle Books, an imprint of Kyle Cathie Ltd.
192–198 Vauxhall Bridge Road
London SW1V 1DX
general.enquiries@kylebooks.com
www.kylebooks.com

10 9 8 7 6 5 4 3 2 1

ISBN 978 0 85783 229 0

Editor: Judith Hannam
Editorial Assistant: Hannah Coughlin
Design: Ketchup
Photographer: Catherine Gratwicke
Prop Stylist: Polly Webb-Wilson
Production: Lisa Pinnell

A Cataloguing in Publication record for this
title is available from the British Library.

Colour reproduction by ALTA London
Printed and bound in Singapore by Tien Wah Press

ACKNOWLEDGEMENTS

This book has been mulling about in my head for
many a year. It is a book of notes and scribbles, trials
and tinkering. It has all come together with the help
of so many wonderful & talented folks who saw the
jumble in my head and heart and helped it evolve
onto the page and out in to the world of folk who are
also keen on walking the simple, uncomplicated life.

A great big thank you to all the folk at Kyle Books,
especially Judith, my editor, who always looks at me
with a twinkle in her eye when I am maybe being
a little bit crackers and makes the most sensible
suggestions to set me back on the right path …

To Kyle, for allowing me to share my forgotten
ways with the world … and to the rest of the team
at Kyle Books who get the show on the road.

To Cath, who yet again is an amazing taker of pictures
and catches all the little things that make sense of a
shot, all the while listening to my ramblings about
complete randomness and useless facts.

To Polly, your magic sprinkles across a surface, drips
from a spoon, a slight twitch to the left or a shuffle
to the right gives life to each and every shot.

And last but not least, my sweet girl and boys.
They have helped me test recipes, swallowed down
various concoctions when they were ill, modelled
various bits & bobs along with sticky in pins and bits
of tape, tried to be quiet when I needed to write and
put up with my mad shouting ways when the block
set in. You three are my everything and make each day
a little sunnier.

LIVE EACH SEASON AS IT PASSES; BREATHE THE
AIR, DRINK THE DRINK, TASTE THE FRUIT AND RESIGN
YOURSELF TO THE INFLUENCES OF EACH.'
HENRY DAVID THOREAU